CHANGE AND MOBILITY IN CONTEMPORARY INDIA

This book studies caste and community dynamics in India and offers a critical view of social mobility from below. Building on the theories of the eminent sociologist M. N. Srinivas, the chapters in this volume reformulate the debate on caste as they document the changing inter-caste dynamics and caste-based violence in contemporary India.

The volume showcases the new language of change in caste relations, articulated mostly from the perspective of the marginalised as experiences, differences, contestations, assertions and as citizenship rights. It focuses on the clash between traditional structures of inequality and the ideals of equality and justice in a liberal, democratic India. It also highlights the persistence of caste and endogamy and the interlocking nature of caste, gender and disability, struggles of ethnic groups and informal workers in the market economy, discrimination in the labour market and the dissolution of dissent in the public sphere.

With contributions from leading scholars of social change and development in India and abroad, this volume will be useful for scholars and researchers of sociology, social anthropology, minority and subaltern studies, and development studies.

Sobin George is Assistant Professor at the Centre for Study of Social Change and Development, Institute for Social and Economic Change, Bengaluru. He is the author of the book *Caste Embeddedness of Rural Public Health Services* and *Work and Health in Informal Economy*. He has also published several research articles in national and international journals.

Manohar Yadav is Professor of Sociology at the Institute for Social and Economic Change, Bengaluru. He has published extensively on issues concerning subaltern and marginalised sections of the society in India. He is an active member of Dalit movements in Karnataka.

Anand Inbanathan is a Sociologist who recently retired from the Institute for Social and Economic Change, Bengaluru. He has been engaged in research on decentralisation and local government and has also worked on health and development as well as on migration. He is the author of the book *Migration and Adaptation: Tamils in Delhi.*

CHANGE AND MOBILITY IN CONTEMPORARY INDIA

Thinking M. N. Srinivas Today

Edited by
Sobin George, Manohar Yadav and Anand Inbanathan

Routledge
Taylor & Francis Group
LONDON AND NEW YORK

First published 2020 by Routledge

2 Park Square, Milton Park, Abingdon, Oxon OX14 4RN
605 Third Avenue, New York, NY 10017

Routledge is an imprint of the Taylor & Francis Group, an informa business

First issued in paperback 2021

British Library Cataloguing-in-Publication Data
A catalogue record for this book is available from the British Library

Library of Congress Cataloging-in-Publication Data
Names: George, Sobin, editor. | Yadav, Manohar S., editor. | Inbanathan, Anand, editor.
Title: Change and mobility in contemporary India : thinking M.N. Srinivas today / edited by Sobin George, Manohar Yadav and Anand Inbanathan.
Description: Abingdon, Oxon ; New York, NY : Routledge, 2020. | Includes bibliographical references and index.
Identifiers: LCCN 2019028899 (print) | LCCN 2019028900 (ebook)
Subjects: LCSH: Social mobility—India. | Caste—India. | Social stratification—India. | Social change—India. | Srinivas, Mysore Narasimhachar.
Classification: LCC HN690.Z9 S65165 2019 (print) | LCC HN690.Z9 (ebook) | DDC 305.5/130954—dc23
LC record available at https://lccn.loc.gov/2019028899
LC ebook record available at https://lccn.loc.gov/2019028900

ISBN: 978-1-138-33430-4 (hbk)
ISBN: 978-1-03-217709-0 (pbk)
DOI: 10.4324/9780429345074

Typeset in Sabon
by Apex CoVantage, LLC

CONTENTS

FIGURES

TABLES

CONTRIBUTORS

Kripa Ananthpur is Associate Professor at the Madras Institute of Development Studies, Tamil Nadu. She focuses on exploring the dynamics of local democracy and governance and the interface between governance and civil society. She has worked extensively on local governance and decentralisation in collaboration with the Institute of Development Studies, Sussex, UK, the World Bank and other national and international organisations.

Tanweer Fazal is Associate Professor at Centre for the Study of Social Systems, School of Social Sciences, Jawaharlal Nehru University, New Delhi. He specialises in political sociology with specific interest in sociology of nationalism and nationhood, minority and ethnic identities and social movements. He is author of the book *"Nation-state" and Minority Rights in India: Comparative Perspectives on Muslim and Sikh Identities*.

Sobin George is Assistant Professor at the Centre for Study of Social Change and Development, Institute for Social and Economic Change, Bangalore. He holds a PhD from the Centre of Social Medicine and Community Health, Jawaharlal Nehru University, New Delhi. His areas of research cover marginalities, sociology of health and illness, bio-political economy of the medical industry and labour rights. He has published extensively on social gradients of health and informality. He is the author of the book *Caste Embeddedness of Rural Public Health Services* and *Work and Health in Informal Economy* as well as editor of *Redefined Labour Spaces: Organising Workers in Post Liberalised India* and *Health System Strengthening: Experiences of Developing Countries*. He has also published several research articles in national and international journals and edited books. He was previously Project Officer at the Centre for Education and Communication, New Delhi (2006–2008), a Fellow at the Indian Institute of Dalit Studies, New Delhi (2008–2010) and Assistant Director at the Indian Council of Social Science Research (ICSSR), New Delhi (2010–2012).

Anand Inbanathan is a Sociologist who recently retired from the Institute for Social and Economic Change, Bengaluru. He holds a PhD in Sociology from the Centre for the Study of Social Systems, School of Social Sciences, Jawaharlal Nehru University, New Delhi. He has been engaged in research on decentralisation and local government for many years and has considerable publications in the area. He has also worked on health and development as well as on migration. He is the author of the book *Migration and Adaptation: Tamils in Delhi.*

Sujit Kumar teaches Political Science at St. Joseph's College, Bengaluru. His research and publications are on adivasi politics, political economy of land, and social movements. His recent research work is related to the political economy of outsourcing in coal mining. Apart from chapters in edited volumes he has published research articles in *Studies in Indian Politics*, *Economic and Political Weekly*, *Seminar* and other journals.

C. M. Lakshmana is Professor and Head of the Population Research Centre at the Institute for Social and Economic Change, Bengaluru. He has a PhD in Geography and his areas of research include gender and health, environment and society and population geography. He has authored six books and published several research articles in national and international journals.

S. Madheswaran is Professor of Economics at the Institute for Social and Economic Change, Bengaluru. He has been engaged in teaching and research in Applied Econometrics for the past 23 years. He has extensively written in the areas of economics of education, economics of labour with special emphasis on discrimination in the labour market, environmental economics and applied econometrics and published in several reputed national and international journals.

James Manor is the Emeka Anyaoku Professor Emeritus of Commonwealth Studies in the School of Advanced Study, University of London. He has taught at Yale, Harvard and Leicester Universities; at the Institute of Development Studies, University of Sussex; and at the Institute for Social and Economic Change, Bengaluru. His latest books include *Politics and State-Society Relations in India;* and *Politics and the Right to Work: India's Mahatma Gandhi National Rural Employment Guarantee Act* (co-authored with Rob Jenkins).

Reimeingam Marchang is Assistant Professor at Institute for Social and Economic Change (Bengaluru). He holds a PhD in Economics from Jawaharlal Nehru University, Delhi. His broad areas of research work include labour issues, migration studies, human resource development, regional economy and regional development.

Abdul Raoof C. K. is a research scholar at Institute for Social and Economic Change, Bangalore, India. He works on questions of dominant discourses and its formation and normalisation in the society.

Smrutirekha Singhari is Assistant Professor at the Amity School of Economics, Noida. She has a PhD in Economics from University of Mysore through the Institute for Social and Economic Change, Bengaluru. Her areas of research include economics of caste discrimination and labour market discrimination.

Joseph Tharamangalam is Emeritus Professor of Sociology at Mount Saint Vincent University and Adjunct Professor of International Development Studies at St. Mary's University, both in Halifax, Canada. His areas of research have included agrarian relations and peasant movements in India, religious pluralism and secularism, and more recently, comparative studies on human development across Indian states and different countries.

B. P. Vani is Associate Professor at the Centre for Economic Studies and Policy (CESP), Institute for Social and Economic Change, Bangalore, India. She holds a Masters in Statistics from Bangalore University. Her research focuses on poverty and income distribution and she has published papers in journals of international repute.

Radhika Viswanathan is a development researcher and development communication specialist. Her interests lie in the intersection between research, policy and communication in the development sector, with a particular interest on local governance and democracy, climate change and gender. She cohosts and produces podcasts on India and development, notably "In the Field" and "Seachange". She holds a Masters degree from the London School of Economics and Political Science and the University of Strasbourg.

Manohar Yadav is Professor of Sociology at the Institute for Social and Economic Change, Bengaluru. He completed his MA, PhD, from the department of Social Anthropology, Karnataka University Dharwad, Karnataka. He served as a founding principal of Dr Babasaheb Ambedkar Centenary Degree College, Hubli from June 1991 to May 1995. During his academic career at ISEC he has carried out several research projects sponsored by different agencies including departments of Government of India, Government of Karnataka, and the World Bank. He has published extensively on issues concerning subaltern and marginalised sections of the society in India. He was a visiting fellow at Cornel University during the year 2005. He is an active member of Dalit movements in Karnataka.

FOREWORD

M. N. Srinivas (1916–1999) was one of the doyens of sociology of his time and a scholar who institutionalised a new perspective to assess change in contemporary India. In the beginning of the 1950s he formulated a question for sociologists to investigate: What is the nature of social change in contemporary India? To answer this question he presented a new conceptual architecture for sociology. To this effect he formulated concepts such as sanskritisation, westernisation and dominant castes giving sociologists a new perspective to perceive the momentous changes occurring in India. In the process he shifted the focus of sociology from an historical-civilisational approach which his first supervisor, G. S. Ghurye had codified, to an assessment of empirical study of contemporary developments through a "field view" approach. Since then the gaze of sociology and sociologists has remained focused on an empirical investigation of contemporary processes that impact the everyday lives of the people of India.

In asking this question and in insisting on the methodology of the "field view", Srinivas was following in the footsteps of European sociologists such as Karl Marx, Emile Durkheim or Max Weber. They analysed the changes occurring as a consequence of the transition to modern society by discussing the formation of new classes, or the role protestant ethic played in the growth of capitalist society, or the way division of labour allows for mechanical solidarity to emerge. And in doing so each of them created a methodology to examine these changes empirically. Srinivas focused his gaze on the way *jati*(s) or caste(s) were becoming mobile as a consequence of the economic, political and social changes being ushered in in newly independent nation-states. In addition to jatis and their dynamics, he was also interested in examining the way the structure of villages was changing and assessing how individuals were transiting towards towns and cities. His text *Rampura* has been frequently quoted not only because of the way it was conceived (from memory) but also for the change it depicts occurring in modern India. Now more than five decades since he formulated this perspective, this volume is reviewing this legacy by once again asking: what is the nature of change in India and does Srinivas's conceptual architecture remain relevant today?

Contemporary social theory across the world has critiqued Marxist, Weberian and Durkheimian perspectives together with those of other late nineteenth- and early twentieth-century sociologists and questioned their current relevance. The same critique has emerged in India with regard to Srinivas. The chapters in this volume should be read in this context and the extensive questioning by Marxist, Dalit and feminist standpoints and by those who argue that Srinivas's nationalist and disciplinary biases did not allow an assessment of the impact of global processes in examining social change in India. Given the implication of global articulations, sociology, they would argue, needs to move away from the legacy of Srinivas.

Srinivas believed that he was living in a very important moment in the history of the country – a moment which he later termed "living in a revolution". Though it was momentous, he suggested that it was a quiet revolution, "bloodless, continuous, progressively more inclusive, and faster". He believed that sociologists and social anthropologists should assess the way the process of post-independent nation building was transforming Indian society. This was done through adult franchise, protective discrimination and land reforms. While this process was non-violent he was also conscious that it might lead to violence in the future and that it is not easy for a country as "large, old, diverse and stratified as India" to change and transition to modernity without violence. "The road to nation-building", he argues, "is a long and hard one" (Srinivas, 1992, p. 28).

Engaging with Srinivas at this juncture can help sociologists ask how to re-craft the sociological language as we assess the contemporary consequences of social change in India seven decades after Srinivas made his position explicit. This is a challenge this book poses to its readers. I hope the chapters in this volume will help scholars re-think ways to debate change in contemporary India.

Sujata Patel,
Indian Institute of Advanced Study, Shimla

Reference

Srinivas, M. N. (1992). *On living in a revolution and other essays* (pp. 13–29). New Delhi: Oxford University Press.

ACKNOWLEDGEMENTS

This edited volume is an outcome of the international seminar organised by the Centre for the Study of Social Change and Development at the Institute for Social and Economic Change (ISEC), Bengaluru in August 2017 as part of the birth centenary year celebration of Professor M. N. Srinivas. We are grateful to the Indian Council of Social Science Research (ICSSR), New Delhi and the Department of Higher Education, Government of Karnataka for generously funding the seminar and the edited volume.

We thank Dr A. Ravindra, IAS, the then Chairman of the Governing Board of ISEC, Bengaluru, who helped us immensely to arrange funding for this project. We thank Professor M. G. Chandrakanth, Director, ISEC, Bengaluru for his encouragements and help. We thank Professor Sujata Patel, who played a key role in the seminar from its inception till completion. We sincerely thank all the authors who wholeheartedly cooperated with us during various stages of the production of the book. Other than the authors here, discussions in the seminar were enriched by eminent social scientists in the country including Professor Shiv Visvanathan, Professor Gopal Guru, Professor N. Jayaram, Professor K. G. Karanth, Professor Valerian Rodrigues, Professor R. S. Deshpande, Professor M. V. Nadkarni, Professor D. Rajasekhar, Professor K. P. Kannan, Professor Supriya Roy Chowdhury, Dr Janaki Abraham, Dr V. Anil Kumar, Professor R. R. Siva Prasad, Professor Kala Sridhar, Professor John S. Moolakkattu and Dr Jos Chathukulam. We thank them all.

We thank the unknown reviewers of this volume. Their meticulous comments and suggestions helped us refine the chapters further. We are grateful to Shoma Choudhury, commissioning editor at Routledge, Shloka Chauhan and the entire editorial team of Routledge for their help in the production of this volume.

We thank all the participants in the seminar, whose comments were of immense help to sharpen the arguments of the chapters in the edited volume. We sincerely acknowledge the support extended by the administration, publication division, ISEC library staff and our student friends Abdul

Raoof, Arun B. Chandran, Anitha Ravindranath, Biplaw Kumar Singh, Geeta Sahu, Jyoti Thankur, Muhammed Saalim, Omkar Nadh, Romica Vasudev and Shreejata Niyogi at ISEC, Bengaluru.

Sobin George, Manohar Yadav and
Anand Inbanathan

ABBREVIATIONS

AIUMM	All India United Muslim Morcha
AKSMTF	Akhila Kerala Swathanthra Matsya Tozhilali Federation
AL	Agriculture labour
Ar.P	Arunachal Pradesh
BCCL	Bharat Coking Coal Ltd.
BJP	Bharatiya Janata Party
BSP	Bahujan Samaj Party
CAG	Comptroller and Auditor General of India
CFS	Centre for Fisheries Studies
CICP	Committee on Integrated Coal Policy
CIL	Coal India Limited
CITU	Centre of Indian Trade Unions
CLA	Contract Labour (Regulation and Abolition) Act
CPI(M)	Communist Party of India (Marxist)
CPRC	Coal Price Revision Committee
CSO	Central Statistics Office
CTUOs	Central Trade Union Offices
DES	Directorate of Economics & Statistics – Government of Manipur
DVC	Damodar Valley Corporation
EBC	Extreme Backward Classes
ECL	Eastern Coalfield Limited
FBOs	Faith based organisations
FC	Forward Caste
GBPIHED	G.B. Pant Institute of Himalayan Environment & Development
GP	Gram Panchayat
HDI	Human Development Index
HMI	Household Manufacturing Industry
ICTs	Information and Communication Technologies
IFC	International Financial Commission
IFPRI	International Food Policy Research Institute
ILO	International Labour Organisation

INTUC	Indian National Trade Union Congress
ISEC	Institute for Social and Economic Change
ISRO	Indian Space Research Organisation
JDU	Janata Dal United
JMM	Jharkhand Mukti Morcha
JMS	Janta Mazdoor Sangh
KLCFF	Kerala Latin Catholic Fishermen Forum
LCC	Latin Catholic Church
MBC	Most Backward Caste
MDO	Mining, Development and Operation
MEF	Ministry of Environment & Forests – Government of India
MNREGA	Mahatma Gandhi National Rural Employment Guarantee Act
MP	Member of Parliament
NCEUS	National Commission for Enterprises in the Unorganised Sector
NCHW	National Council for Handicapped Welfare
NDA	National Democratic Alliance
NE states	North Eastern states
NER	North Eastern region
NFF	National Fishermen's Forum
NGT	National Green Tribunal
NHFDC	National Handicapped Finance & Development Corporation
NICDR	National Information Centre on Disability & Rehabilitation
NSSO	National Sample Survey Office
NTUI	New Trade Union Initiative
OBC	Other Backward Castes
OCPs	Open Cast Projects
OL	Other Labour
PDOs	Panchayat Development Officers
PM	Post-Mandal
PMM	Pasmanda Muslim movement
PPP	Public Private Partnership
PSU	Public Sector Unit
RGCCI	Registrar General and Census Commissioner of India
RLBs	Rural Local Bodies
RRTC	Regional Rehabilitation Training Centre
RTI	Right to Information
SDMC	School Development and Management Society
SNDP	*Sree Narayana Dharma Paripalana Yogam.* Translated as "Organisation to propagate the teachings of Sree Narayana Guru"
STs	Scheduled Tribes
TCS	Transnational Civil Society
TISCO	Tata Iron and Steel Company

TMF	Trivandrum Matsyathozhilai Federation
TWU	Tata Workers' Union
UDF	United Democratic Front
UNDP	United Nations Development Program
Vi-MAC	Vizhinjam Mother Port Action Council
VISL	Vizhinjam International Seaport Limited
VVES	Viyapari Vyavasayi Ekopana Samithi
WHO	World Health Organisation
WPR	Work participation rate
YMCA	Young Men's Christian Association

GLOSSARY

Adi-Dravidas A term used by the state of Tamil Nadu in India to denote Dalits since 1914
Adivasi A member of any of the aboriginal tribal peoples living in India before the arrival of the Aryans in the second millennium BC
Ajlaf Local converts
Anganwadies A type of rural child care centre in India
Arzal Converts to Islam from ex-untouchable Hindu castes
Ashraf Persons claiming descent from prophet Muhammad/foreign conquerors
Bahujan Literally meaning "People in majority"
Beda (in Kannada) "Don't want/need"
Beedi Mini-cigar filled with tobacco flake and commonly wrapped in a leaf tied with a string or adhesive at one end
Bhaya Fear
Biradari Fraternity
Bunyaadi Basic
Chalvadi That troubles, torments, annoys, harasses
Chamars A Scheduled Caste community in India traditionally considered outside the Hindu ritual ranking system of castes
Chambhar Chambhar/Charamkar stands for one who creates leather products from *chamda* (leather).
Chettiyars Chettiar or Chetti is a title used by many mercantile, agricultural and land-owning castes in Tamil Nadu and Kerala
Dalit Sena Literally means "Dalit army"
Dhaba Roadside restaurant in India
Dharmashastras A genre of Sanskrit texts, and refers to the treatises (shastras) of Hinduism on dharma.
Dhor A Scheduled Caste in Maharashtra, India
Dwijas Literally means "twice-born". The concept is premised on the belief that a person has physical and spiritual births.
Galli The narrowest lane of the city

Gram Panchayats Local self-governance system in India at the village or small-town level

Gram sabhas Village council

Gram Swaraj Literally means self-governance or "self-rule" of the village

Haath Means "hand" in English

Harijans Term popularised by Mohandas Gandhi for referring to communities traditionally called "Untouchable"

Hartals A type of strike/protest by closing shops, halting public and private transport

Holeya A Scheduled Caste of India, mainly belonging to present day Karnataka State, Kerala, Maharashtra, Tamil Nadu and also Madhya Pradesh

Jagirdars The feudal lords of the Jagirdari system that existed in South Asia during the thirteenth century, wherein the power to govern and collect tax from an estate was granted to an appointee of the state

Jai Bhim Dal Victory to Bhim (referred to Dr. Ambedkar)

Jamiat-ul Ansar Literally meaning organisation of Ansar (weaver caste among Muslims)

Jamiat-ul Quresh Translated as "organisation of Quresh" (butchers and meat sellers)

Janakeeya Kootaima People's collective

Samrakshana Samithi Translated as "People's Collective Council for Protection"

Jat A traditionally agricultural community native to the Indian subcontinent who are primarily of Hindu, Muslim and Sikh faiths

Jati Referred to group of clans, tribes, communities, and sub-communities, and religions in India

Jolaha One who weaves, or whose occupation is to weave

Kallar caste One of the three related castes of southern India that constitute the Mukkulathor confederacy, along with the Maravar and Agamudayar; constitute a united social caste on the basis of parallel professions in the Tamil Nadu region

Kriya Vedike Translated as "Karnataka Dalit Action forum"

Sangarsha Samithi Translated as "Organisation for Dalit agitations in Karnataka"

Janandolan Sangha Translated as "Karnataka peoples collective" Khutba Sermon, Sermon

Kurubas A shepherd caste in Karnataka

Madiga A Dalit artisan caste group, who live mainly in the southern states of India: Andhra Pradesh, Kerala, Tamil Nadu, Karnataka

Mala Dalits caste group from the south Indian states of Andhra Pradesh, Telangana and Karnataka. Mala groups are considered as Scheduled Castes by the Government of India.

Mangs Dalit caste mainly residing in the state of Maharashtra
Nawab Term for a native governor during the time of the Mogul empire
Nawabiyat Nawabhood
Naya neta Literally means "new leader"
Pallars A peasant caste community found in the Indian state of Tamil Nadu and parts of Sri Lanka; they are traditionally tenant agriculturists
Paraiyars A caste group in Tamil Nadu, India and Sri Lanka who are also known as Adi Dravida
Muslim Mahaz Marginalised Muslim Front
Pradhans A notable surname used mostly by the Brahmin, Kshatriya Gurjar, and Kayastha people of northern, western, and southern India
Qaum A community of people who share a common language, culture, ethnicity, descent, and/or history
Raj Rule; government
Rig Veda An ancient Indian collection of Vedic Sanskrit hymns along with associated commentaries on liturgy, ritual and mystical exegesis
Ryotwari One of the two main systems during British rule in India used to collect revenues from the cultivators of agricultural land
Sahyaparvatha Western Ghats
Samrakshana Samithi Translated as Council for Protection of Western Ghats
Samudhayam Literally means "community"
Santhal adivasis An ethnic group, native to Nepal and the Indian states of Jharkhand, Assam, West Bengal, Bihar and Odisha
Sarpanch The head of a village elected by the village-level constitutional body of local self-government called the Panchayat in India
Shivalingam An abstract or aniconic representation of the Hindu deity Shiva in Shaivism
Siyaasat Literally means "Politics"
Swathanthra Literally means "Independent"
Udaiyar Also known as Udayar, a caste found in the state of Tamil Nadu, India. They are composed of various sub-groups such as Srutiman, Nathaman, Malayaman, Moopanar and Nainar
Upanishads Ancient Sanskrit texts that contain some of the central philosophical concepts and ideas of Hinduism
Vanniyar A community classified as Shudra in the Hindu varna system
Varna A Sanskrit word which means type, order, colour or class which refers to social classes in Brahminical books like the *Manusmriti*
Zamindari The system under which zamindars held land and the right to collect tax on behalf of imperial courts or for military purposes
Zulm hua hai Literally means "injustice has happened"

M. N. SRINIVAS AND THE KALEIDOSCOPE OF SOCIAL CHANGE IN MODERN INDIA

A critical introduction

Sobin George, Manohar Yadav and Anand Inbanathan

Indian Sociology and Anthropology have produced several legendary scholars, with a widespread academic following throughout the world. Among them, Professor M. N. Srinivas stands out for his outstanding contributions. He is essentially known as the doyen of Indian Sociology and Social Anthropology. Rather, he has evolved as a legendary scholar. M. N. Srinivas was born at a point of time when Indian society was passing through radical changes. The changes brought about in the fields of law, education, communication, transport, etc. had turned out to be sudden invasions on the deep-rooted historical structures of India. The conservative essence of Indian society was only steadily yielding to accept and adopt the new pervasive character of these severe changes. Naturally, as he grew as a scholar, M. N. Srinivas became increasingly preoccupied with understanding the nature of these changes. Born in a Brahmin family, and growing up in that insular setting, Srinivas himself was struggling to understand his own neighbourhood. In fact, as he was steadily exposed to the immediate world around him, he experienced many cultural shocks.

Unsurprisingly, M. N. Srinivas's areas of academic interest covered changes that had been occurring in villages, culture, caste, and social relations as a whole (Srinivas, 1956; Srinivas, 1966; Srinivas, 1994). His book *Social Change in Modern India* published in 1966 (a compilation of lectures), posed some of the fundamental questions related to social change, its direction, orientations and sources of orientations, which still inform us on the trajectories of social change in India. His conceptualisation of social change through concepts such as *westernisation* in order to make sense of the process of modernisation begun by colonial rulers in India, and *sanskritisation* to explain the upward mobility of groups lower in the caste hierarchy through imitating the rituals of higher castes, which made it possible for a lower caste to attain a higher level in the hierarchy, is the subject of these lectures, and has enlightened a vast number of people, scholars and others, about the manner in which groups seek to find a better, if not a higher, place for themselves. The studies on Rampura, the Karnataka village, where he carried out field work for several years, produced another important concept, *dominant caste*, which helped to explain the process through which middle-level caste groups become socially, economically and politically powerful. The relevance of the illustrious concepts such as sanskritisation, dominant caste and vote bank that he developed is still not lost; although social changes have necessitated scholars to critically appreciate them. In fact, the concepts that he had developed were the foundations of serious scholarly thought and debate, which were set in motion through his writings. As a scholar and teacher, Srinivas has had an enormous influence on sociology and anthropology in India, and the study of Indian society.

Critical engagements on the writings of M. N. Srinivas are also necessary at least on a few significant counts. Srinivas himself was open to criticism and was well aware and appreciative of the sweeping changes brought about by greater processes such as industrialisation, modernisation and above all, what he termed as the democratic revolution. He was often criticised for his confinements to the disciplinary boundaries of western Anthropology and Sociology to understand Indian realities and the difficulties of his change thesis in situating some of the non-changing/less changed characters of caste in its core. Another criticism is that for a substantial part of his writings, M. N. Srinivas relied significantly on field work, which in his view was the only way to produce authentic and reliable material for anthropological studies and perhaps he over-relied on what he termed "participant observation". However, the truth is that research in sociology and anthropology, like any other social science discipline, has undergone transformations and became accustomed to other epistemological approaches and methods of enquiries influenced by critical traditions, post-colonial theories and subaltern studies. Methods like small ethnographies, visual ethnographies, survey research, lived experiences, experiential epistemologies, oral histories, reflexive traditions, discourse analysis and increasing use of secondary resources and big data also have become acceptable in sociological research like in other social science disciplines. Of course, among sociologists and social anthropologists, lengthy ethnographic field work still holds significant influence. However, the practice of staying several months at a time in the field is slowly waning away.

M. N. Srinivas and the contemporary questions of social change

How relevant are Srinivas and his conceptual categories of social change today? This is indeed a complex question. Contemporary Indian society is changing at an unprecedented pace, mostly triggered by the developments in the real world, related to advancements in science and technology, the emergence of new world systems and governance paradigms, regional integration, war and conflicts, new forms of dominance and resistance, ecological imbalances and above all, the ways in which systems and life-worlds respond to such developments. Social science academia has been engaging with several questions which have emerged as part of larger processes of change. For instance, sociological studies for a significant period of time had deeply engaged with the changes in the village economy and agrarian relations in India. While there are context specific understandings and explanations available from such studies, several questions still emanate from the contemporary processes of change. What are the new relationships that have come up in the rural village economy with the loosening of caste ties, and what is the nature of the emergent power relations? Have such changes disintegrated the structures of oppression and humiliation?

Rural transformation and mobility of lower caste groups

One of the major settings of caste and a key factor for understanding the changing nature of caste relations in Indian society has been the village, mainly due to its caste-centred agricultural production relations, high dependency of the lower caste groups on the primary and allied sectors and the replication of such dependency and related caste relations in other civic spheres in one form or the other. (It should also be noted that caste has been an important socio-cultural marker in urban India as well, especially in the spheres of housing and private and public employment.) Locating caste in agriculture and related sectors also laid a special emphasis on changing agricultural relations and patterns in the studies on the changing nature of caste. The obvious assumption was that caste is changing with the loosening of *jajmani*, patron–client and other forms of dependency and the long-term contract relations of lower castes with the landed castes, mainly the upper and middle-level caste groups who are moving to urban areas (Gupta, 2005; Jodhka, 2012). Studies following this pattern have therefore argued that a major change in caste relations occurred in at least some parts of rural India.

It is generally agreed that the characteristics of caste have shown resilience in several spheres of village life, primarily due to structural changes in the village economy, social and political mobilisations, identity assertion movements and affirmative action policies of the state. However, it is important to understand whether such changes are making meaningful transformations in caste relations, and most importantly, their continuing disadvantages with respect to other social groups. Scholars have made substantial efforts to study the mobility of different caste groups in rural India to understand the pattern of changes that have evolved over a period of time.[1] There is general agreement that lower castes, especially some caste groups among Dalits who benefited from reservation policies and had a relatively better socio-economic origin, moved upward both socially and economically (Ram, 1988). It is also highlighted that overall, occupational ranking has become less rigid, as Mayer (1996) has argued. Similarly, studies showed that there is a trend among Dalits who are endowed with skills for modern occupations and professional education moving up as they increase in number (McMillan, 2005). While some of these studies noted the upward mobility of Dalits, they also highlighted the increasing inter-group disparities in such movements or, in other words, the "worsening of their relative disadvantage" (ibid.: 147) in the labour market and other spheres. Ashwini Deshpande (2011), for instance, noted that the Dalits who moved out of traditional occupations by breaking caste ties were usually stuck at the lower rung of the economic class, highlighting that such movement did not translate into the betterment of their economic condition. The reasons are several, ranging from higher initial levels of vulnerabilities and discrimination (Madheswaran & Attewal, 2007) to lack of social capital and effective business network support (Jodhka, 2012; Harriss-White et al., 2014) and economic exclusion (Harriss-White et al., 2014).

Others, such as Mayer (1996), argued that caste is still identified with occupation in India. George (2018), drawing from the recent data from the NSSO survey, has highlighted the share of Dalits in various occupational categories to understand the dimensions of such mobility. He noted that nearly 53 per cent of Dalits were in the sectors of agriculture, fisheries and forestry in rural India in 2011–2012, followed by construction (21 per cent). While manufacturing accounts for 8 per cent of the total Dalit workers in rural India, there are around 5 per cent Dalits in wholesale and retail sectors. These figures confirm that even though they tend to move out of agricultural sectors, it is mostly to highly informal jobs, making them equally or more economically vulnerable due to the precariousness associated with non-formal employment. Since such changes do not endow them with economic mobility, the social mobility associated with moving out of agriculture sector is in question.

The availability of public works in rural India, more specifically the employment generation programme of the Mahatma Gandhi National Rural Employment Guarantee Act (MNREGA), which aims to ensure up to 100 days of work for members from poor households, is often projected not only as an avenue for coming out of poverty but also from caste ties for Dalits. This programme, arguably, provided more opportunities for Dalits to find employment outside the agricultural lands of middle- and upper-level caste groups. Scholars argue that MNREGA has brought about several important changes such as reduction of the male–female wage gap, increase in rural wages and reduction in distress migration (Chandrasekhar & Ghosh, 2011). Have such programmes helped Dalits to move economically upward? Though no evidence is available on the economic mobility of Dalits due to MNREGA, studies have highlighted that they participate in large numbers in these programmes. For instance, the India Human Development Survey (2004–2005) by the National Council of Applied Economic Research (NCAER) and the University of Maryland found that 36 per cent of Dalits and 30 per cent of Adivasis participated in the programme, which is much higher than their population share. Though their share decreased over time, the trend of overrepresentation of Dalits and Adivasis was visible in 2011–2012 as well (Desai & Reeve, 2012). The increased participation by Dalits in employment guarantee programmes leads to several propositions. First, it reiterates the continuing economic deprivation of Dalits in rural India. It questions the economic mobility thesis even if one argues that it is mostly the women and the elderly who are seeking work under MNREGA. Second, it shows that if employment is available outside agriculture, nothing (caste ties) significantly stops them from participating in it. But a decision to do so also means dropping the possibility of work that they might get otherwise with the landed castes. This supports the "change" thesis of caste, especially its social mobility part. The third is that though there is a favourable environment existing for social mobility for Dalits and in most instances they could move out of the older forms of bondages, loyalties and submissions to upper- and middle-level castes, questions on whether available economic opportunities such as MNREGA and others are adequate to enable social mobility remain. For instance, the uncertainties in assured employment under MNREGA and the informalities in other sectors can bring them back to the agriculture fields of land owners, and thus limit the possibility for a sustained social and economic mobility.

Caste and modern occupations

The recent discourses on the mobility of Dalits centre on economic liberalisation and the "new opportunities" created in the non-conventional spheres. While some scholarship draws attention to the opportunities of the "new and caste-neutral occupations" available for Dalits (Kapur, Prasad, Pritchett, & Babu, 2010; Kapur, Babu, & Prasad, 2014), others argue that Dalits are "unfavourably included" (Teltumbde, 2008; Thorat, Mahamallick, & Sadana, 2010) in such markets and the market principles and ideologies could not "break down India's caste-based social order, but reinforce it in some ways" (Harriss-White et al., 2014). It is important to understand these two arguments a little more carefully for their potential to direct research on the mobility of lower castes in this way, which could either strengthen or undermine the arguments of neoliberalism as an ideology. Whether the new opportunities created by the two decade–long processes of economic liberalisation are caste-neutral is one of the crucial questions relevant here. Though the available literature does not clearly indicate the caste characteristics of such new opportunities, they talk about their nature (mainly the capital–labour relations) in considerable detail.

The arguments in favour of the linkages of economic liberalisation, new opportunities and upward mobility of Dalits are mainly formed around two occupational categories – self-employment or entrepreneurship and jobs – which include the salaried class of professionals and lower- and middle-level employees in the private service and manufacturing sectors. Such literature gives a fairly optimistic picture of the fluidification of caste with more Dalits entering into entrepreneurship ranging from small- and medium- to large-scale industrial sectors to "escape their predicament" (Kapur et al., 2014). The question, however, is whether there is a noticeable wave of Dalit entrepreneurship in India other than the stories one hears from some pockets in states like Rajasthan, western Utter Pradesh and Delhi. Iyer, Khanna, and Varshney (2011), based on data drawn from economic censuses up to 2005, argued that Dalits and Adivasis were "significantly underrepresented" in the ownership of enterprises. One should also read this along with the data from the NSSO 68th round on employment and unemployment, which showed that among Dalits, as much as 51 per cent are in elementary occupations that include casual labourers, agricultural workers, street vendors, garbage collectors and domestic help (see George, 2014). The study by Iyer et al. (2011) highlighted that on average, the firms owned by Dalits and Adivasis are smaller than those owned by other social groups. This means that their establishments are susceptible to easy external shocks resulting in closures due to big retailers or other market dynamics. Another major observation they made is that most of the firms run by Dalits did not have access to outside financial sources and were without the necessary links to the market, which would significantly limit their ability to grow. Hence, they argue that the projected wave of an affluent Dalit entrepreneurship is not valid. Also, there is no substantial evidence to link it to the opening up of the economy and most importantly to non-discrimination based on caste. However, there is evidence that in some places, as much as 70 per cent of the firms of Dalit entrepreneurs are within their caste occupation such as sanitary ware, washing clothes, leather works, barbershops, wood works, groceries, etc. (Harriss-White et al., 2014).

Discussions on economic liberalisation as an impediment and opportunity in general are also relevant for the lower caste groups, especially the Dalits in India. The general contingencies of economic liberalisation in the initial years along with the measures of macroeconomic stabilisation and structural adjustment programmes have had significant implications for the welfare provisions of the people in the country. The post-liberalisation period has witnessed several structural changes in the labour market such as the rapid expansion of the informal sector, informalisation of the formal sector, flexible production systems that emphasise efficiency and labour productivity, stagnation of wages, growth without employment generation, reduction of public expenditure in sectors like education and health and the involvement of the market in provisioning of services such as drinking water and electricity, to mention a few. There is ample literature on how such processes deprived the poor, who mostly consist of marginalised sections of Dalits and Adivasis (Teltumbde, 2001). Some scholars are of the view that liberalisation had a "twofold impact" on Dalits – the shrinking of formal sector jobs in the public sector, which has been a huge opportunity for them for secure employment (ibid.), and a higher entry barrier due to discrimination in private sector jobs, which have come up as part of the thesis of new opportunities (Jodhka, 2010). Harriss-White et al. (2014) further added that the market has always been in tandem with the social and economic hierarchy of the caste system either directly or indirectly. Thorat et al. (2010) also confirmed that the occupations of Dalits in the liberalised economy predominantly remained in the lowest rung of the occupational hierarchy or the same caste-based occupations such as casual labourers, agricultural workers, sanitation workers, carcass cleaners and so on. Discrimination against Dalits and minorities, especially Muslims, in the entry levels of private sector job markets was discussed in detail by several scholars in the book edited by Thorat and Neuman (2012).

Viewing changes from below: articulation of marginalities and development deficits

There is general agreement that M. N. Srinivas always adopted a top-down approach to understanding many of the social phenomena, especially caste relations, that he chose to analyse. It is also true that not all of Srinivas's ideas and theorisation received complete acceptance – perhaps due to this approach. To be precise, he always looked at things from the viewpoint of the upper- and middle-level caste groups. For instance, his concept of dominant caste is not adequate to understand the mobility of Dalits in India. Hence, some outcomes of his research as knowledge were always questionable, to some extent, in their corresponding value as the "reality". And to that extent, his writings cannot be treated as fully flawless. In fact, M. N. Srinivas himself agreed with what he termed as his own shortcomings (Tharamangalam & Chathukulam, 2018).

However, of late, we have seen several studies that transcended these conceptual barriers and looked at the mobility of socio-religious groups from the standpoint of the lower caste groups, who articulate it through their experiences with caste in day-to-day life (Guru, 2016). Some of these studies also looked at it from the standpoint of caste as a barrier to equal participation in society, as a development category (Chakraborty, 2003; Hasan & Mehta, 2006; Shah, Mander, Thorat, Deshpande, & Baviskar, 2006; Desai & Dubey, 2012; Holmes, Sadana, & Rath, 2010; Rizvi, 2011; Borooah, Sabharwal, & Naik, 2013), as an identity and agency for political mobilisation (Manor, 2012) and as an effective pressure group to pursue group specific development needs (Jodhka & Sirari, 2012; George, 2012). These scholarships, interestingly, did not often invoke the concepts of sanskritisation or dominant caste to understand the mobility of ex-untouchable castes, the discrimination that they faced and their development deficit. This was a new language of caste mostly articulated from the perspectives of the marginalised as experience, as differences, as contestations, as assertions and as citizenship rights.

The exclusion of social groups on various markers of marginalities is an important thrust area, which some of these studies addressed (Deshpande, 2000; Nambissan, 2009; Thorat & Newman, 2007; Jodhka, 2010; Acharya, 2010). Who are the social, religious and ethnic groups that have been suffering from development deficit due to exclusion? What are the forms of discrimination? What are their consequences and impact? How do they articulate their rights and entitlements? What are its gender dimensions? Other important forces of change, which are discussed in this literature, relate to the new social and political movements and their increasing role in catalysing the processes of change. Recent movements such as the Dalit movements, backward class movements for reservation, movements of minorities for group-specific policies, anti-corporate movements, movements for autonomy and self-determination, movements for land, anti-corruption movements, etc. are important examples. They evoke questions such as what is the nature of such movements and how have they evolved? What are the issues that such movements address? Can such movements act as effective pressure groups to influence policy making? What are their roles in the political processes of the respective regions? How do the people who are represented by these movements perceive the role of such movements in improving their material conditions?

The changing role of the state, governance and political participation of people are another set of relevant questions in the studies of the directions of social change (Inbanathan & Sivanna, 2010; Martin, 2015). Is democratic decentralisation an inclusive process? Are programmes implemented at the field level with the active participation of the beneficiaries? Can we say whether the existing structures of power relations limit the participation of certain groups of people in such programmes? How do beneficiaries see such programmes?

The resurgence of the questions of religion, especially through the lens of development and citizenship also pointed to the direction of social and economic change. Among others, the questions of cultural nationalism and minoratisation assume paramount importance. Sachar Committee report and reports of minority commissions on the differences within minority religious groups, especially Muslims and Christians also brought about new dimensions of citizenship rights to the development deficits of minorities. It is now accepted that caste is a reality among Muslims (Ahmad, 2003;

Sikhand, 2004) and Christians (Lobo, 2001; Oommen, 2010) in India. However, the practice of caste varies in form and content across localities and regions. Available data on the social and economic conditions of Dalit Christians shows that they are worse off than Dalit Sikhs in poverty levels (Deshpande & Bapna, 2009). The inter-caste differential of poverty is reported to be the highest among Christians since the upper caste Christians are much better off than Dalit Christians. Dalit Christians are comparatively better off than Dalit Muslims, Dalit Sikhs and Buddhists in terms of education and occupation, but worse off than non-Dalit groups. The report on the Social, Economic and Educational Status of the Muslim Community of India (2006) by the Justice Rajendra Sachar Committee revealed their poor status in terms of education, poverty, and representation in jobs and other labour markets. For instance, the report has revealed that the literacy level among Muslims is vastly below the national average. It is also striking that 25 per cent of children of Muslim parents in the 6–14-year age group have either never attended school or have dropped out. With regard to the availability of educational infrastructure, the report revealed that there is a clear and significant inverse association between the proportion of the Muslim population and availability of educational infrastructure in small villages. It is also important to note that access to government schools for children of Muslim parents is limited. Data based on various rounds of NSSO reveals the levels of poverty among Muslims. An inter-religious comparison of poverty shows that in rural areas, poverty was the highest amongst the Buddhist population followed by Muslims and Hindus and in urban areas, the highest among Muslims followed by Buddhists and Hindus. Occupational patterns and representation in public sector jobs of Muslims show that Muslim-OBCs are significantly deprived in comparison to Hindu-OBCs. The work participation rate (WPR) shows a sharp difference between Hindu-OBCs and Muslims. The share of Muslim-OBCs in government/PSU jobs is much lower than that of Hindu-OBCs.

Similarly, we have seen articulations of change, continuity and mobility of ethnic groups, exploited workers and women through the lens of citizenship rights. For instance, studies have made fairly good attempts to attribute the continuing impoverished conditions of Adivasis in India to factors such as discrimination and isolation (Jayaraj & Subramanian, 1999; Borooah, 2005) and deprivation in terms of access to productive income earning assets (Thorat & Mahamallik, 2007) and utilisation of available resources such as land and common property resources (Gang, Sen, & Yun, 2008). Some studies highlighted the persisting impoverishment of Adivasis in the context of lack of equal opportunities and treatment. For instance, Borooah (2005) highlighted that "at least one-third of the average income/probability differences between Hindu and SC/ST households was due to the 'unequal treatment' of the latter". Some studies brought out the interconnectedness of lack of education and gainful employment to the higher level of poverty among Adivasis. For instance, Gang et al. (2008) showed how lack of educational attainment and gainful occupation resulted in the poverty of Adivasis in India.

The new language of marginalities and development was also pronounced in studies that looked at changes like the increasing fragmentation of the organised sector, breakdown of the traditional rural village economy and growing informalisation, which have resulted in new forms of employment and labour relations, mostly in the informal sectors. Studies have highlighted that poverty showed an increasing trend in the households of casual workers in urban India, and households of agriculture workers in rural areas, and most importantly the households of Dalits and Adivasis who form the highest share of casual labour (Sen & Himanshu, 2004; Thorat, 2010). This raised several questions such as how working in the informal sector and being poor are associated; what the dimensions of gender differentials of informal employment and poverty are; and how people perceive their relative position of poverty and what the coping mechanisms to come out of it are.

Indian society has undergone several changes in its traditional social structures, institutions, culture and belief systems in the recent past. The social science academia has contributed significantly to the understanding of the diverse processes of these changes, their dimensions, trajectories and policy imperatives. The contribution of Professor Srinivas to understand these multifaceted and complex social processes has been immense. Most of the conceptual categories that he developed are still relevant and being widely used by scholars in India and outside to make sense of the processes of social mobility and change in caste, religious and ethnic communities. We have also seen meaningful critical engagements on these categories developed by Srinivas by social scientists to explain several initiatives for change from the people who are at the lower end of the social hierarchy. The chapters in this volume critically engage with some of the complex questions of changes, continuities and mutations of caste groups, which they confront in their social and economic mobility. The purpose of this book is to locate these multiple trajectories of social change in modern India from various standpoints.

The book is organised into three sections. Chapters in the first section deal largely with the questions of various contemporary explanations of changing caste relations in social and political spheres that include differences, continuities, flexibilities, contestations, conflicts, accommodation and mutations of caste characteristics. James Manor discusses the incremental changes in caste relations in rural India that are moving from refusal of caste hierarchies by the lower castes and resultant violence to "negotiated accommodation" in rural India. He further argues that in most of the regions he studied, accommodations and stalemates tend to predominate over violence. He further explains the processes of these changes elaborately in the chapter. Manohar Yadav critically analyses the *ostensible* changes in caste relations and the mobility of lower castes that are often attributed to sanskritisation. Emphasising on the flexibilities and adaptive nature of caste and its manifestations in caste relations, he argues that mobility through processes like sanskritisation is not fully real when changes are viewed from the standpoints of the lower castes. Anand Inbanathan offers an in-depth discussion on the hierarchical caste relations between the "dominant lower castes" and Dalits that are reflected in the local self-governments in Tamil Nadu villages. He argues that caste has not changed much in its essence in rural Tamil Nadu and the Dalit identity still takes precedence over any other entitlement such as education or wealth that the person may possess. Kripa Ananthpur and Radhika Viswanathan, however, provide an account of substantial change in caste relations towards contestations and assertions of Dalits against caste hierarchies in rural Karnataka. They argue that with the opportunities provided through the local panchayats, Dalits in northern Karnataka villages could loosen the caste-based traditional structure of dominance and patronage networks and pressurise the traditional political elites to adapt and accommodate.

The chapters in the second section deal with questions of mobility from the perspective of lower caste groups and their strategies for the same. Joseph Tharamangalam critically examines the field view of M. N. Srinivas, which he thinks has a top-down approach and argues that articulations from below have significantly changed the understanding of the trajectory of caste mobility of the lower caste groups. Sobin George argues that caste relations are situational in nature since caste has come out of structures to accommodate itself to spaces and situations. Drawing from an empirical study conducted in a northern Karnataka village, he highlights that there was no common story of change in caste relations for Dalits even in a micro space and a single person, family or sub-caste can signify several views of caste, which makes the familiar explanations of caste relations based on hierarchy, differences, flexibility, conflict, accommodation or any other across spaces and situations problematic when looked at from below. Drawing from the Pasmanda movement in Bihar and Maharashtra, Tanveer Fazal sheds light on the caste-like subdivisions within Islam, their development deficit and initiatives for resistance and recognition from below.

Chapters in the third section broadly deal with the questions of marginalities and development, of those who are left behind in the process of *mainstream* development. S. Madheswaran, Smrutirekha Singhari and B. P. Vani examine the extent of caste-based discrimination between young and older cohorts using national-level representative data spanning from the period 1983 to 2011–2012. They find that the effect of discrimination is acute among (Dalit) young workers, which is a major barrier for their social and economic upward mobility. C. M. Lakshmana attempts to situate "disability" in the wider canvass of social stratification, especially the caste division in Indian society and examines the upward mobility of persons with disability across social groups in India. Reimeingam Marchang examines the changing livelihood system from agricultural to non-agricultural of the Scheduled Tribes (STs) of the North Eastern region of India. He associates the social and economic mobility of various tribal groups in North East India to their gradual movement from agricultural to non-agricultural occupation with educational advancement in general, although there are several markers of deprivations still persisting in the region. Sujit Kumar examines how practices like outsourcing of organised sector industries informalise labour, which is already marginalised on other markers such as ethnicity. Drawing from the coalmines of Dhanbad, Jharkhand, he illustrates how informalised labour relations can intertwine with local feudal establishments (he refers to mafia politics here) with the decline of strong labour movements and multiply the vulnerabilities of workers. Abdul Raoof C. K., in his chapter, invites attention to what is happening to the legitimate resistance movement of civil society organisations of the marginalised communities against neoliberal development projects of the state. Drawing from the case of a sea port project in Kerala, he illustrates how a progressive faith-based organisation, which stood up to protect the livelihood of project affected traditional fishermen, is appropriated to change their stand to support the project and use their moral authority over people to build consent.

We hope that the chapters in this volume will be useful to the students of social change in India.

Note

1 For a comprehensive review, please see Vaid (2014)

References

Acharya, S. S. (2010). Caste and patterns of discrimination in rural public healthcare Services. In S. Thorat & K. S. Newman (Eds.), *Blocked by caste: Economic discrimination in modern India* (pp. 208–229). New Delhi: Oxford University Press.

Ahmad, I. (2003). A different Jihad: Dalit Muslims challenge to Ashraf hegemony. *Economic and Political Weekly*, *38*(46), 4886–4891.

Borooah, V. K. (2005). Caste, inequality, and poverty in India. *Review of Development Economics*, *9*(3), 399–414.

Borooah, V. K., Sabharwal, N. S., & Naik, A. K. (2013). *Caste, employment, and wages in India: How do employees from different social groups fare in India's labour market?* New Delhi: Indian Institute of Dalit Studies.

Chakraborty, U. (2003). *Gendering caste through a feminist lens*. Calcutta: Popular Prakashan.

Chandrasekhar, C. P., & Ghosh, J. (2011). Public works and wages in rural India. *Macroscan*. Retrieved from www.nregaconsortium.in/download/articles/jayati%20and%20chandrashekhar%2011-jan-11.pdf

Desai, S., & Dubey, A. (2012). Caste in 21st century India: Competing narratives. *Economic and Political Weekly*, *46*(11), 40–49.

Desai, S., & Reeve, V. (2012). *India Human Development Survey-II*. Ann Arbor, MI: Inter-University Consortium for Political and Social Research. doi:10.3886/ICPSR36151.v2

Deshpande, A. (2000). Does caste still define disparity? A look at inequality in Kerala, India. *The American Economic Review*, *90*(2), 322–325.

Deshpande, A. (2011). *The grammar of caste: Economic discrimination in contemporary India*. New Delhi: Oxford University Press.

Deshpande, S., & Bapna, G. (2009). *Dalits in Muslim and Christian communities: A status report on current social scientific knowledge*. New Delhi: Nation Commission for Minorities, Government of India.

Gang, I. N., Sen, K., & Yun, M.-S. (2008). Poverty in rural India: Caste and tribe. *Review of Income and Wealth*, *54*(1), 50–70. Blackwell Publishing.

George, S. (2012). *Discrimination, development deficit and the question for group-specific policies* (Working Paper Series, Vol. VI). New Delhi: Indian Institute of Dalit Studies.

George, S. (2014). India's retail trade revolution: Socio-religious dimensions of employment loss and new forms of urban exclusion. *Journal of Exclusion Studies*, *4*(2), 191–203.

George, S. (2018). *The caste-embeddedness of rural public health services* (ISEC Monograph 59). Bengaluru: Institute for Social and Economic Change.

Gupta, D. (2005). Whither the Indian village: Culture and agriculture in rural India. *Economic and Political Weekly*, *40*(08), 751–758.

Guru, G. (2016). Shifting categories in the discourse on caste and class. *Economic and Political Weekly*, *L1*(47), 21–25.

Harriss-White, B., Basile, E., Dixit, A., Joddar, P., Prakash, A., & Vidyarthee, K. (2014). *Dalits and Adivasis in India's business economy: Three essays and an atlas*. New Delhi: Three Essays.

Hasan, R., & Mehta, A. (2006). Under-representation of disadvantaged classes in colleges: What do the data tell us? *Economic and Political Weekly*, *41*(35), 3791–3796.

Holmes, R., Sadana, N., & Rath, S. (2010). *Gendered risks, poverty and vulnerability in India: Case study of the Indian Mahatma Gandhi National Rural Employment Guarantee Act (Madhya Pradesh)*. New Delhi: Overseas Development Institute and Indian Institute of Dalit Studies.

Inbanathan, A., & Sivanna, N. (2010). *Scheduled castes, legitimacy and local governance: Continuing social exclusion in panchayats* (Working Paper, No. 257). Bangalore: Institute for Social and Economic Change.

Iyer, L., Khanna, T., & Varshney, A. (2011). *Caste and entrepreneurship in India* (Harvard Business School Working Paper). Retrieved from www.hbs.edu/faculty/Publication%20Files/12-028_0e6e69ba-da4d-4abc-b66a-9611dc07c53b.pdf

Jayaraj, D., & Subramanian, S. (1999). Poverty and discrimination: Measurement and evidence from rural India. In B. Harriss-White & S. Subramaniam (Eds.), *Ill-fare in India: Essays on India's social sector in honour of S Guhan* (pp. 196–226). New Delhi: Sage Publications.

Jodhka, S. S. (2010). Dalits in business: Self-employed scheduled castes in North-West India. *Economic and Political Weekly*, *45*(11), 41–48.

Jodhka, S. S. (2012). Agrarian changes in the times of (neo-liberal) 'crises': Revisiting attached labour in Haryana. *Economic and Political Weekly*, *47*(26–27), 5–13.

Jodhka, S., & Sirari, T. (2012). *In the footsteps of Ambedkar: Mobility, identity and Dalit initiatives for change* (Working Paper Series, Vol. 6, No. 1). New Delhi: Institute for Dalit Studies.

Kapur, D., Babu, D. S., & Prasad, C. B. (2014). *Defying the odds: The rise of Dalit entrepreneurs*. New Delhi: Random House.

Kapur, D., Prasad, C. B., Pritchett, L., & Babu, D. S. (2010). Rethinking inequality: Dalits in Uttar Pradesh in the market reform era. *Economic and Political Weekly*, *45*(35), 39–49.

Lobo, L. (2001). Visions, illusions and dilemmas of Dalit Christians in India. In S. R. Charsley & G. K. Karanth (Eds.), *Cultural subordination and the Dalit challenge*. New Delhi: Sage Publications.

Madheswaran, S., & Attewal, P. (2007). Caste discrimination in the Indian urban labour market: Evidence from the National Sample Survey. *Economic and Political Weekly*, *42*(41), 4146–4153.

Manor, J. (2012). Accommodation and conflict. *Seminar*, *633*, 14–18.

Martin, N. (2015). Rural elites and the limits of scheduled caste assertiveness in rural Malwa, Punjab. *Economic and Political Weekly*, *50*(52), 37–44.

Mayer, A. (1996). Caste in an Indian village: Change and continuity 1954–1992. In C. J. Fuller (Ed.), *Caste today* (pp. 32–64). New Delhi: Oxford University Press.

McMillan, A. (2005). *Standing at the margins: Representation and electoral reservation in India*. New Delhi: Oxford University Press.

Nambissan, G. B. (2009). *Exclusion and discrimination in schools: Experiences of Dalit children*. New Delhi: Indian Institute of Dalit Studies and UNICEF.

Oommen, T. K. (2010). India's religious minorities and state policy. In R. R. Patil & J. Dabhi (Eds.), *Dalit Christians in India* (pp. 17–39). New Delhi: Manak Publications.

Ram, N. (1988). *Mobile scheduled castes: Rise of a new middle class*. New Delhi: South Asia Books.

Rizvi, F. F. (2011). *Housing situation among the poor and marginalised rural households: A study of Indira Awaas Yojana in selected districts of Orissa and Maharashtra*. New Delhi: Indian Institute of Dalit Studies.

Sen, A., & Himanshu (2004). Poverty and inequality in India: II: Widening disparities during the 1990s. *Economic and Political Weekly*, 4361–4375.

Shah, G., Mander, H., Thorat, S., Deshpande, S., & Baviskar, A. (2006). *Untouchability in rural India*. New Delhi: Sage Publications.

Sikhand, Y. (2004). *Islam, caste and Dalit-Muslim relations in India*. New Delhi: Global Media Publications.

Srinivas, M. N. (1956). A note on sanskritization and westernization. *The Journal of Asian Studies*, *15*(04), 481–496.

Srinivas, M. N. (1966). *Social change in modern India*. Berkeley, CA: University of California Press.

Srinivas, M. N. (1994). *The dominant caste and other essays*. New York: Oxford University Press.

Teltumbde, A. (2001). *Globalisation and the Dalits*. Nagpur: Sanket Prakashan.

Teltumbde, A. (2008). *Khairlanji: A strange and bitter crop*. New Delhi: Navayana.

Tharamangalam, J., & Chathukulam, J. (2018). Revisiting the legacy of M N Srinivas. *Economic and Political Weekly*. Retrieved June 18, 2018, from www.epw.in/engage/article/revisiting-legacy-m-n-srinivas

Thorat, A. (2010). Ethnicity, caste and religion: Implications for poverty outcomes. *Economic and Political Weekly*, *45*(51), 47–53.

Thorat, S., & Neuman, K. S. (2012). *Blocked by caste: Economic discrimination in modern India*. New Delhi: Oxford University Press.

Thorat, S., & Newman, K. S. (2007). Caste and economic discrimination: Causes, consequences and remedies. *Economic and Political Weekly*, *42*(41), 4121–4124.

Thorat, S., Mahamallick, M., & Sadana, N. (2010). Caste system and pattern of discrimination in rural markets. In S. Thorat & K. S. Newman (Eds.), *Blocked by caste: Economic discrimination in modern India* (pp. 148–177). New Delhi: Oxford University Press.

Thorat, S., & Mahamallik, M. (2007). *Human development and the status of social groups in Gujarat* (IDS Working Paper Series, Vol. I, No. 6). New Delhi: Indian Institute of Dalit Studies.

Vaid, D. (2014). Caste in contemporary India: Flexibility and persistence. *Annual Review of Sociology*, *40*(1), 391–410.

Part I

CASTE TODAY

Resilience, conflicts and contestations

1

SOME DEBATABLE ARGUMENTS ABOUT CHANGING INTER-CASTE DYNAMICS IN RURAL INDIA[1]

James Manor

This chapter presents several debatable arguments that have emerged from a research project. The arguments in this chapter should be treated as tentative and in need of critical examination. The research project analyses the implications of the increasing (and increasingly obvious) refusals by the so-called "lower" castes – especially Dalits – to accept caste hierarchies in rural areas. (To say this is not to argue that caste – or more specifically as that word is used here, *jati* or endogamous caste group – has weakened. It has not. It is the power and acceptance of caste hierarchies that has waned, not the institution of jati.) What follows from these refusals? Inter-caste tensions have certainly increased as a result – a change which was inevitable, and which is not entirely a negative development.[2] But do we see increases in violence (and at times, more severe violence than 20 or 30 years ago), or grudgingly negotiated accommodations between castes to curtail or pre-empt violence, or uneasy stalemates between castes which stop short of violence?

After hundreds of interviews with villagers, social scientists and other knowledgeable informants in 12 diverse regions and sub-regions of India since 2012 – and after a careful sifting of textual and quantitative evidence – it appears that all three of these things have increased in recent years. But in most (although not all) regions, accommodations appear to have outnumbered acts of violence – by which I mean physical violence (against persons and property). Those accommodations result not from a change of heart among the so-called "higher" castes, but from a change of mind.[3] There is next to no evidence of empathy among them for Dalits. Their self-restraint amid inter-caste tensions is the result of calculations that violence will ensnare "higher" castes in unacceptably severe difficulties.

Because this is a study of inter-caste dynamics, it will focus on each of the three groups: Dalits, the dominant (or in many areas, the formerly dominant) landed castes, and Other Backward Classes (OBCs). To keep the analysis manageable, urban areas are excluded, and so are *adivasis* or Scheduled Tribes. Let us now consider several arguments, all of which are debatable.

Violence between castes has increased, but so have accommodations and stalemates

It is impossible on present evidence to develop precise measurements of changing levels of physical violence between castes. The available numbers show increases, but it is not possible to determine exactly how much of this is explained by more effective reporting. Nevertheless, it appears that violent incidents between castes have been on the rise in recent years (Manor, 2017).[4] However, it is important to stress two qualifying findings from field research. First, when inter-caste tensions arise in villages, there have also been increases in the numbers of negotiated accommodations and of uneasy stalemates between castes, both of which end or pre-empt violence. Second, accommodations appear to outnumber violent incidents in most (but not all) regions and sub-regions of India.[5] Thus, all three of these phenomena – inter-caste violence, accommodations and stalemates – have occurred more often in recent years. This is largely explained by increases in inter-caste tensions as the so-called "lower" castes increasingly (and more openly) refuse to accept caste hierarchies.

Inter-caste violence within one village almost never spreads to other localities

When communal violence (violence between religious groups) occurs – overwhelmingly in urban areas (Varshney, 2002)[6] – there is always a danger that it may spread to other places (including other parts of the same urban centre). By contrast, inter-caste violence in rural areas almost never spreads from one village to another. In over one hundred interviews, people of all caste backgrounds living in localities near villages where serious inter-caste violence had taken place expressed not caste-based anger, resentment or blame, but surprise and often incredulity about those clashes.

How can this be explained? The tensions, disputes and perceived provocations that lead to inter-caste violence in any individual village acquire their meaning and their destructive potential from specificities that are usually distinctive to that village. They seldom have much resonance in nearby villages. In many cases, this is partly explained by marked variations in the social composition of different villages. For an example of these – and there are a great many others – consider the variations that Adrian Mayer found in two Madhya Pradesh villages where he did his field research (Table 1.1).

Thus, if clashes occur between members of different jatis in one village, those specific jatis are often not major players in nearby localities. And even when the social composition of nearby villages differs little from the strife-torn

Table 1.1 Caste rank by population in Madhya Pradesh villages

Caste	*Dewas Senior*	*Ramkheri*
Rajput	1	2
Weaver	2	4
Tanner	3	5
Brahman	4	9
Farmer	5	1

Source: Mayer (1960, p. 41).

village, the outbreak of violence in the affected village is usually triggered by quarrels or incidents not between entire jatis, but between individuals, households or very small groups – by disputes which have no counterparts in nearby villages. So, to reiterate, most inter-caste violence in villages originates from very distinctive local specificities which are not found elsewhere. There are rare exceptions, in areas where political parties or external organisations break through those specificities and stoke inter-caste tensions from without. That can generate severe conflicts, but it is – so far – unusual.[7]

Interviews with villagers also indicated that violent incidents in a nearby village are seen – when they are happening, and in retrospect – as extremely regrettable because they get in the way of more important preoccupations. They are viewed as cautionary episodes which remind those living nearby of the importance of avoiding such clashes. I was repeatedly told by members of landed castes that if such violence were to occur in their villages, the clashes themselves and then the unwelcome interventions from external actors that were likely to follow would disrupt their struggles to make a decent living from agriculture, to educate their children and – because it would mar the reputation of their entire village – to find them good marriages. Dalits and other poorer groups stressed that violence would disrupt their efforts to scratch a living as labourers, as cultivators of tiny plots of land[8] or as small-scale entrepreneurs, and to get their children basic education. Clashes between castes are understood – across caste lines – to threaten the aspirations, interests and livelihoods of diverse groups.

Finally, in most villages, there were enough exits, distractions and disengagements from village politics by the so-called "higher" castes – and in many cases enough exits for Dalits – to prevent people from unduly exacerbating inter-caste tensions within their localities. (See the discussion below of exits, distractions and disengagement.)

Understanding inter-caste accommodations

Thirty years or so ago, acts of violence by landowning castes against Dalits were often (i) measured, and based on (ii) calculations by (iii) elders or senior leaders within the "higher" castes. To understand how, in recent years, inter-caste accommodations have become more common than spasms of lethal violence, we need to consider all three of these things – and how they have changed.

When accommodations are forged, senior figures within the landed castes nearly always still play key roles (except in villages where OBCs are more important protagonists), along with Dalit counterparts (although as we shall see below, the authority of both sets of elders within their castes is sometimes in doubt and under threat). In most recent episodes, decisions by members of the landed castes to opt for negotiations and accommodations – that is, for restrained actions – are based on careful tactical calculations by them. Crucially, those decisions are not inspired by a new-found empathy for Dalits among "higher" castes – which I found to be almost totally lacking. Instead, they are the results of a change of mind and not of heart.

What types of calculations are made? Senior figures within the landed castes know that any violent act, lethal or otherwise, is very likely to become known in the world beyond their village – for several reasons. Dalit organisations are often afflicted by factional disputes, but in most parts of India, they penetrate at least tenuously into many rural areas. Thus, they usually learn of violent incidents, and the result may be a visit to the village by eminent urban Dalits to investigate and publicise the episode. Other (non-Dalit) civil society organisations which are sympathetic to Dalits may also call attention to a violent event. The media now penetrate fairly effectively into rural arenas as well, so that they often report such incidents. When such reports surface, most (but not all) state governments often respond as well – because most ruling parties cannot afford to alienate Dalit voters through inaction. Judges may also enter the picture – especially if petitions are filed by public interest lawyers, a common occurrence. And behind all of these potential intrusions into the village looms the Atrocities Act of 1989.[9]

That Act is a draconian law which characterises a broad array of actions – many of which stop short of physical violence – as "atrocities" against Dalits. It empowers the authorities to jail alleged offenders immediately and without bail, on the strength of an accusation that an atrocity has taken place. It places the burden of proof on the accused rather than the accuser.

State governments vary in the zeal with which they implement the Act. Many are less than assiduous in doing so.[10] Even when a state government seeks to make robust use of the Act, conviction rates for alleged offenders tend to be quite low. But that should not blind us to the fact that even if a member of a "higher" caste is confident about escaping conviction, s/he (usually "he") faces a prolonged legal struggle – often lasting years – which is expensive and excruciatingly inconvenient. It imposes huge opportunity costs since it impedes the accused from getting on with other things that are mightily important: cultivating lands, marketing agricultural produce, developing other enterprises that may yield income, arranging loans, overseeing investments, getting children properly educated and married off, etc. For these reasons, even though convictions are unlikely at the end of the legal process, members of the "higher" castes often view the Atrocities Act with trepidation. In 1999, this writer heard clear expressions of anxiety from members of landowning castes when the Act was discussed in the Gwalior region of Madhya Pradesh where their social/political dominance and the old hierarchies had undergone little erosion.

All of these considerations weigh on the minds of leaders of the landed castes in most villages. They take them into account when deciding how to proceed when inter-caste tensions rise, often because some sort of non-violent incident which they regard as an affront has occurred. These worries are usually sufficient to persuade them to seek to negotiate accommodations with Dalit leaders. Those accommodations are usually uneasy and grudging, but they (and not violence) represent the predominant trend in most regions. What types of 'incidents' occur? There is space here only for inadequate comments on this complex, delicate topic. But let us consider three types of incidents.

Accommodations within villages are sometimes facilitated by a perception among most members of all castes that certain incidents – quarrels and modest ructions, however angry, between members of "higher" castes and Dalits – are rather minor events that do not threaten the prestige of either group. They are not seen by most members of the "higher" castes to have violated old caste norms. This is often true of disputes which appear to involve unusually truculent individuals or households. They are perceived to be rather isolated, private squabbles that do little to disrupt the equanimity of social relations more broadly.

Another set of incidents, which are deeply serious, often trigger not violent conflict but frantic efforts to undo or otherwise resolve them – and sometimes to conceal them. That is often the response when amorous entanglements develop between individuals from Dalit and "higher" castes, when those individuals elope, or when rapes occur. If these occurrences become known beyond the village, *all* of the young women there will be viewed by outsiders with suspicion (unfairly of course) that loose morals prevail in the village, and/or that young women there may have been molested. Those suspicions will make it difficult to find husbands for all of those young women.

Such incidents may trigger violence, but that is not necessarily the most common outcome. Lest suspicions engulf all young women in the village, the first reaction by elders within it is not to resort to violence (which will only call attention to the embarrassment) but to hush things up. If an amorous relationship has developed between members of different castes, efforts are made to separate the two people, and perhaps to buy their silence. If an elopement has occurred, attempts are made to bring the pair back to the village, to annul a marriage if it has taken place, and again to separate the two and to use threats and /or enticements to break their links.

Rapes, which are obviously acts of violence in themselves, are even more inflammatory events. They require more extensive discussion than is possible here. In some parts of India, members of the landed castes have long routinely made free use of the bodies of Dalit women, and this vile practice persists in some areas. But now that Dalit and other organisations and the media have extended their reach into many rural areas, such outrages are more likely to become known – and action under the Atrocities Act may follow. Given that possibility, and given the importance of protecting the reputations of all young women in the village, efforts are again often made to hush things up, to intimidate those who might reveal that it has happened, and perhaps to compensate the victim.

Finally, let us consider a further set of events which are not trivial squabbles between individuals or households. They are not as salacious as amorous relationships, elopements and rapes, but they pose genuine threats to peace in villages because they are perceived by members of the "higher" castes as serious insults to their dignity, and to what remains of the old hierarchical order. By way of illustration, consider two such incidents, both of which occurred in similar areas of South India but which produced contrasting outcomes.

In the first, a few young Dalit men attended a village cinema. They had earned enough money to pay for the most expensive seats – for "chairs" when the cheaper alternatives were "benches" and the "floor". During the screening, they placed their feet on the backs of the seats in the row in front of them – a row in which young women from the local landed caste were sitting off to one side. Such things are seen in India as a patently insulting act. When word spread among members of the landed caste in the village, they were infuriated, and a few began violent attacks on Dalits which soon mushroomed into a killing spree that left many Dalits dead. Criminal investigations ensued, and several members of the landed castes were eventually convicted and jailed for long periods.

In the second case, three Dalits boys teased a lone boy from the landed caste which had formerly dominated social life in the village. When this was reported to landed caste elders, they gathered to consider what action to take. Some advocated violence, to teach the three Dalits a lesson. But the more influential elders argued that such action would produce too many lasting problems for the members of their caste, so they organised a meeting with Dalit elders. After some negotiation, it was agreed that the three Dalit boys would apologise to the boy whom they had teased. After that happened, the incident was considered by all to be closed. The kind of accommodation that was reached in the latter case is far more typical than the savage outbreak of violence in the first.

Accommodations are usually negotiated between senior figures from different castes, between elders. They share one common interest. No matter which caste they belong to, they see themselves as custodians of some sort of predictable "order" in the village (or what is left of it after recent changes) – an "order" within which they enjoy influence and often pre-eminence. Violence would threaten not just tranquillity but predictability – which is of great value to all of those senior figures – and their pre-eminence.

Violence can threaten predictability in several ways. Quite apart from the injuries that it would (unpredictably) cause, violence could cause lasting disruption to daily village routines and make them less predictable. It might create longer-term resentments that would poison predictable inter-caste relations, and thus the outwardly peaceable "order" within the village, the maintenance of which is seen by many as the responsibility of senior leaders within different castes. Since that would impair their capacity to fulfil that responsibility, it would undermine their authority – within their castes, and in general.

Violence might also invite unpredictable interventions in village affairs by outsiders – by civil society organisations, the media, the police or other government actors at higher levels – undermining the already limited autonomy of the village, autonomy which makes local affairs predictable. Even if no such interventions occurred, a spasm of violence would hand the initiative to those who engaged in it – that is, to persons who were at least temporarily beyond the control of local leaders. That might open the way either to challenges by rivals or to a loss of discipline and coherence within their castes. For all of these reasons, the elders within different castes have compelling reasons to seek accommodations when inter-caste tensions rise. If it is impossible to reach accommodations, their efforts may at least achieve uneasy stalemates which stop short of violence.

It is important, however, to note two further points. One thing that in urban centres has helped to prevent violence is missing in villages. In his study of communal violence, Ashutosh Varshney has demonstrated that in cities where "associational forms of engagement" are "robust" they "promote peace", and "their absence or weakness opens up space for communal violence" (Varshney, 2002, pp. 3, 10, 25–26). In villages, we find few – or usually no – "associational" ties across caste lines. So when inter-caste tensions arise, this resource which exists in some cities is unavailable.

Second, as noted above, when inter-caste accommodations are made during tense times, leaders within "higher" castes in villages act not out of empathy with Dalits, but after calculating what best serves their interests. Similar calculations lead many of them to treat Dalits and other disadvantaged groups, day in and day out, with a minimal degree of outward civility. This entails some gains for Dalits. One leading Dalit social scientist who is not a naïve optimist has described this trend as "the beginnings of a reasonably civilised society".[11] But other perceptive analyses of these changes indicate that they mask efforts by "higher" castes to give only limited ground, and to sustain their advantages and hierarchical relationships in new ways (Waghmore, 2013; Waghmore, 2016). However much greater civility may represent a "moral gain", it is sought to be legitimated by the very social forces that are inimical to that gain.[12] Their civility is double-edged.

Three types of violent inter-caste incidents: a spectrum

Incidents of violence in which members of different castes are involved fall into three broad types which can be arranged in a spectrum. Caste is an element in all three, but it is of far less importance at the two extremes. At one extreme are (1) incidents that are – and are seen by villagers of diverse social backgrounds to be – mainly conflicts between individuals or households. At the other extreme are (3) systemic acts of violence that are usually collective actions, calculatedly organised: conflicts over the socio-economic order in which *class* issues loom large, or conflicts between *external social organisations* or political *parties*. At both extremes, caste colours the conflicts, but it is seldom the predominant element. It is appropriate to call incidents in the middle range between these extremes (2) "inter-caste" conflicts because caste plays the most important role in them.

In recent years, a large majority of violent incidents involving members of different castes have occurred in that middle range – more than at either extreme. For that reason, and because caste looms largest in the middle range, that second type of conflict is the main topic in the analysis that will appear elsewhere, although the others also receive attention.

Note that violent incidents which occur at different points on this spectrum usually differ in the following ways:

Type 1: individual, spontaneous actions
Type 2: collective, spontaneous actions
Type 3: collective, pre-planned actions

One more important variation across these three types can be summed up as follows:

Type 1: inter-caste accommodations are usually unnecessary
Type 2: inter-caste accommodations are usually necessary and achievable
Type 3: inter-caste accommodations are usually unachievable

This requires some explanation. When conflicts and violent incidents between individuals or households (Type 1) occur, inter-caste accommodations are usually unnecessary because those conflicts do not mushroom into collective conflicts between castes – since no caste norm is seen to have been violated. They do not trigger collective reactions from affronted castes. (When incidents involving individuals or households are seen to break a caste norm, then the conflicts tend to become collective "caste" conflicts, so that they belong in Type 2, the middle range of the spectrum.[13])

When violence occurs as a result of systemic conflicts (Type 3) – conflicts over the socio-economic order (Sushmita, 2014),[14] or conflicts spurred by political parties or organisations – accommodations are usually impossible. Both kinds of conflicts are in part influenced by powerful external groups/forces. Elders from different castes within a village (who are usually the main negotiators of accommodations) almost always lack the capacity to resolve those much larger issues and to ease the tensions that arise from them – and they are often disinclined to do so.

However, in the middle range between these extremes (Type 2 on this spectrum) – where the great majority of collective acts of inter-caste violence occur or might occur – there is a need for accommodations. Elders and leaders within conflicting castes often (but not always) possess the willingness, the capacity and the authority within their castes to make them achievable and sustainable (to win their acceptance among members of their castes).

One last word is in order here. If we ask whether any one of the three types of violence is becoming less or more common, the evidence suggests a decline in the frequency of incidents in Type 3 which are substantially class-based – which have occurred, for example, in Bihar. That trend is making Type 2 – inter-caste violence which is the main focus of this study – still more predominant than it has recently been. Readers may wonder whether this claim is undermined by the persistence (and in some regions, the increase) of so-called "Naxalite" violence. But while it fits into Type 3, the groups involved are often adivasis or "tribals" who fall outside this analysis of inter-caste conflict.

It should be added, however, that incidents under Type 3 which are not class-based, but which are triggered from outside villages by social organisations or political parties – for example, conflicts in parts of Tamil Nadu or in Uttar Pradesh since the recent state election – are on the increase. Those incidents often straddle the line between Type 3 and Type 2.

Conditions which increase or decrease the likelihood of collective violence

Let us consider two sets of conditions within certain villages which qualify as opposite extremes. Both sets of conditions can be found in multiple regions of India, and despite the radical contrasts between these conditions, they have one thing in common: they tend strongly not to produce collective violence between "higher" castes and Dalits (which is almost always to say attacks on the latter). Instead, it is in less extreme situations which occupy the middle ground between these extremes that collective violence tends – albeit rather unusually – to occur. This represents a second "spectrum" which focuses on types of villages.

There are, first, villages where "higher" castes and Dalits are both present but lead almost entirely separate lives in separate settlements – which are near to one another, but between which there is very little interaction. Dalits do little or no work of any description for or amid "higher" castes. They no longer perform degrading tasks associated with traditional "sanitation" – removing human waste or the carcasses of dead animals. They do not work on the lands of "higher" castes. (Landowners may employ machines or labourers from elsewhere.) Dalits have their own separate and invariably inferior sources of water, and if their children attend schools, these are distinct from the schools of other castes in the village. By whatever means, they eke out their livelihoods outside the employ of higher castes. There are thus next to no bonds of dependence, or inter-dependence (Srinivas, 2003), between Dalits and "higher" castes. Personal or collective ties – and face-to-face interactions – between individuals or groups from the two adjacent settlements scarcely exist. Members of "higher" castes may not even know the names of most of their Dalit neighbours – and vice versa.

The second set of conditions differs markedly. Dalits and "higher" castes may occupy distinct sections of a village, but they interact often, in diverse ways, so that there is a rather dense web of ties between individuals and groups on each side of the divide. Dalits may still perform some of their traditional tasks, and many of them work as agricultural labourers in the fields of "higher", landed castes. The livelihoods of many Dalits are thus linked to their engagement with "higher" castes and with the village as a whole socio-economic entity. Dalit children may attend the same schools as many children from other castes – although more prosperous "higher" caste families may send their children to private schools in or near the village. Despite increased education, literacy and other benefits and opportunities provided by governments, Dalits are thus still linked to a significant extent by bonds of dependency upon "higher" castes.

But the "higher" castes – whether they recognise the fact or not – are also somewhat dependent upon their Dalit neighbours. (Members of "higher" castes are uncomfortable with this recognition.) They may, often not incidentally, depend on Dalits to perform certain roles in village rituals and festivities. Members of "higher" castes who aspire to leadership posts on *gram panchayats* (which at least since 2005 thanks to the MGNREGA, control substantial funds provided by higher levels of government) may need the votes of Dalits and other disadvantaged groups to win elections. So, there is also a degree of inter-dependence connecting the two groups. Personal and collective ties – and face-to-face interactions – between individuals and groups from the Dalit and "higher" castes occur routinely.[15] Dalits tend to know the names of individuals in the "higher" castes. The reverse is also true – although in only some of these villages do they actually call individual Dalits by their names.

The second set of villages has obviously undergone less social and economic change than the first. In the second set, there is – despite serious inequalities and, often, persisting humiliations for Dalits – some social cohesion (of an invidious type). Any given village in the first set represents a society that is almost entirely lacking in cohesion. In both sets of villages, there is greater predictability in caste interactions than in villages which lie between these two extremes. In the first set, such interactions – predictably – barely occur. In the second set, frequent dealings between "higher" castes and Dalits generate informal expectations and tacit understandings about how interactions are likely to go. And since people on both sides tend to conform to these expectations, there are few surprises – few departures from informal norms. Such surprises and departures in less predictable settings that lie between these two extremes are potential sources of tension and, on occasion, collective violence.

In other ways, however, the very low levels of collective caste violence in these two extreme sets of cases have different explanations. In the first set of villages, since there are only minimal interactions between Dalits and "higher" castes, issues that might trigger violence seldom arise. In the second, more cohesive villages, inter-dependence – whether or not it is clearly recognised by "higher" castes – imposes at least some minimal restraint upon their behaviour towards Dalits. And the patronising treatment which "higher" castes mete out to Dalits carries with it an implicit acknowledgement that some tenuous bonds of mutuality exist, alongside the still substantial dependence of the latter upon the former.

Exits, distractions and disengagements from village power dynamics

One last theme needs attention here. Some members of the formerly dominant landed castes have available to them exits and/or distractions from village power dynamics, and they may choose to disengage from local politics which they find frustrating. Exits include part-time or full-time migrations, usually to urban centres. If they remain in the village, some of them may be distracted by events in urban centres where they have financial investments, properties and second homes, and where their children may have settled. These distractions make them less concerned with village affairs, and less inclined to deal aggressively with Dalits in order to sustain old hierarchical relations there. Others from the landed castes may disengage with local political struggles, often within panchayats, because they view the contestation and cross-examinations that come with such struggles to be vexing and beneath their dignity. All of these things may ease tensions within villages, make village power dynamics less inequitable, and work to the advantage of other castes – including both Dalits and OBCs.

Exits are also available to many Dalits, to travel to find work elsewhere. In some cases, well-paid day labour, often on construction sites, is available in booming urban centres which are near enough to permit commuting – so that heads of households are not absentees from their villages, and families are not uprooted.[16] The reasonably high wages from such work give those Dalits considerable economic autonomy and resources to use for better nutrition, housing, health care and education for their children. In many Indian states, the MGNREGA provided Dalits and other poor people with similar, more modest but far from trivial material gains. Some were able to disengage from invidious bonds of dependency (Jenkins & Manor, 2017).[17] Dalits have derived significant advantages from these things.

The positive potential of exits, distractions and disengagements to make power dynamics in villages less inequitable (or at least more open), and to ease inter-caste tensions, is apparent from Surinder Jodhka's work on localities where exits are not available. In such places, inter-caste relations tend to be more toxic, the possibility of violence is greater, and accommodations are more difficult to arrange.

But exits, distractions and disengagements have double-edged implications. There is negative potential here too. When landed caste leaders exit, become distracted and/or disengage from village politics, their jatis may be left without senior figures who have the authority within their castes to restrain "hot heads" among their caste fellows, to negotiate accommodations, and to make those agreements stick. When they exit or disengage, other less prosperous members of their jatis may feel abandoned and vulnerable. That may lead them to take their anxieties out on Dalits, to resist inter-caste accommodations and acts of minimal civility – and to engage in anti-Dalit abuse and violence. Exits and disengagements by landed caste leaders may create opportunities for OBCs to exercise more power within village politics. Their anxieties about social status are often more acute than among the landed castes. So tensions and even clashes with Dalits may become more likely, and accommodations with Dalits may prove more difficult to arrange. Exits and disengagements by landed caste leaders usually often make social interactions less ordered and predictable – and that sometimes proves destructive.

When Dalits resort to distress migration for long periods to undertake labour elsewhere, they may be forced into demeaning work in vile conditions. They may take their families with them, which disrupts their children's education and may entail the children becoming labourers. Their prolonged absences erode the numerical strength of Dalits within the village. When, by contrast, Dalits commute to nearby urban centres for better paid work, their earnings may inspire jealousy among other castes – similar to the jealousy that sometimes arises when some Dalits take Ambedkar's emphasis on education seriously and make greater progress on that front than some other castes – or when some Dalits make gains thanks to reservations that are not available to others.

For the most part, exits, distractions and disengagements with village politics often make a constructive contribution to inter-caste relations within villages. But in more than a few cases, they have the opposite effect.

Inter-caste accommodations may become more difficult to arrange

Accommodations negotiated by senior leaders of different castes are only enforceable if each leader has the authority within his[18] caste to persuade or compel his caste fellows to accept them.[19] But in many localities, that authority faces the threat or the reality of erosion – as a result of several different processes.

Many of the changes which have contributed to the declining acceptance of hierarchies by members of "lower" castes also threaten the authority of elders within castes. Education, occupational differentiation, and spatial and economic mobility may make individuals more materially autonomous and less deferential to the more senior members of their castes. The emergence of local political entrepreneurs and of enterprising politicians on elected local councils (gram panchayats) may also undermine the pre-eminence of caste elders. The problems that afflict agriculture, and the sagging prestige attached to farming, may weaken the influence of senior figures within landed castes.

Elders' authority is also sometimes undermined by actions which they themselves have taken. This is especially true of leaders of the landed castes. Three themes, noted above, matter here: disengagements, exits and distractions. Some senior figures within the once dominant landed castes have disengaged from panchayat politics after becoming exasperated with the challenges and cross-questioning that they face there from opponents. Surinder Jodhka has shown that such senior leaders may invest funds in urban centres, and establish second homes there for themselves or their children, where they spend part of their time. Even if they do not undertake part-time exits from the village, these things become distractions which prevent them from giving great attention to affairs within their villages and their castes to maintain their influence there.[20]

These and other trends erode the authority structures within castes. On the evidence collected during field work on this project, it appears that in most villages, caste elders have so far managed to minimise the damage. But if the erosion goes much further, it will become more difficult to arrange accommodations between castes, and to enforce compliance once they are arranged. So, the trend towards greater accommodation may not be sustainable. Inter-caste violence may become more common, and the recent predominance of accommodation over violence may not endure.

Notes

1 This study would never have been undertaken had M. N. Srinivas not generously provided this writer with many hours of informal tutorials on Indian society over more than a quarter-century. His kindness to a student who mainly studied politics was a great, astonishing gift.

2 When tensions lead to violence, it causes serious damage and suffering. But it is also an indicator of change that holds some promise of less unjust social relations. Violent incidents also often spur Dalits to greater awareness. I am grateful to S. Japhet for stressing the latter point.
3 I am grateful to K. C. Suri for initially alerting me to this key point.
4 A detailed argument for this claim will appear in the book on this topic. It should be stressed that studies of levels of various *other* types of violence have in the past shown that, despite claims in the media, they tend *not* to follow a linear path, upward or downward. The numbers rise and fall in no discernible pattern. (See for example, the discussion of findings developed by Philip Oldenburg in Manor, J. [2017]. "Ethnicity" in Indian Politics, chapter eight. In J. Manor [Ed.], *Politics and state-society relations in India*. New Delhi/London/New York: Orient BlackSwan/Hurst/Oxford University Press.) But statistics from recent years strongly suggest more or less steadily rising levels of inter-caste violence.
5 Once again, a detailed argument in support of that claim will be proved in the book on this topic.
6 Ashutosh Varshney has written, "Between 1950 and 1995, rural India . . . accounted for less than 4 per cent of the deaths in communal violence. Hindu-Muslim violence is primarily an urban phenomenon." For details, see Varshney (2002, p. 6).
7 Those exceptions will be assessed in detail in the book based on this research. See also the discussion elsewhere in this chapter of a spectrum of types of violent inter-caste incidents. Also note that most political parties lack the organisational capacity to penetrate very effectively into villages. And those that do tend to damp down inter-caste conflict. In a small (and shrinking) number of regions, parties of the left have penetrative capacity, and they seek to restrain inter-caste conflict. In other regions it is the Hindu right, that is, the BJP whose penetrative capacity in rural areas is often quite limited (see Manor, J. [2007]. In part a myth: The BJP's organisational strength, chapter seven. In J. Manor [Ed.], *Politics and state-society relations*) plus the RSS which often supplements that party's organisation – reaches some villages. But these organisations usually seek to build solidarity among Hindu castes, at the expense of minorities. Recall Prime Minister Modi's sharp criticism of *gau rakshaks* who beat Dalits (whose support he had been cultivating) in the Una incident. The BJP currently controls numerous state governments, and in many but not all of those states, its efforts to cultivate Dalit voters prevent it from stoking inter-caste tensions.
8 As Mihir Shah has noted, roughly one-half of poor households in India possess such plots.
9 Its full title is the "Scheduled Castes and Scheduled Tribes (Prevention of Atrocities) Act, 1989".
10 See for example the cover story in *Frontline*, 21 November–4 December 2009.
11 Private communication with this writer.
12 I am grateful to Sasheej Hegde for this point.
13 I am grateful to Kripa Ananthpur for stressing this point.
14 I am not in total agreement with the interpretation, but it provides useful insights into this type of conflict.
15 I am grateful to Sobin George for stressing this point.
16 I conducted interviews while travelling with a group of Dalit villagers from near Nanjangud in Karnataka on their daily commute by train to Mysore city where such work was available.
17 This case is made in great detail in Jenkins, R., & Manor, J. (2017). *Politics and the right to work: India's Mahatma Gandhi National Rural Employment*

Guarantee Act. New Delhi/London/New York: Orient BlackSwan/Hurst/Oxford University Press) especially in chapter six.

18 The leaders are almost always men.

19 The low level of inter-caste violence – and especially of lethal incidents – in some parts of India (for example, the Udaipur region of Rajasthan) is not explained by an absence either of social change or of inter-caste tensions. Both things are present, although tensions are lower than in some other regions. Violence remains low because elders, local leaders among the landed castes and Dalits, both continue to have significant influence and can manage tensions. I am grateful to Sanjay Lodha for stressing this point.

20 I found evidence to corroborate Jodhka's finding in numerous other regions. It emerged in especially rich detail in Andhra Pradesh.

References

Jenkins, R., & Manor, J. (2017). *Politics and the right to work: India's Mahatma Gandhi National Rural Employment Guarantee Act*. New Delhi/London/New York: Orient BlackSwan/Hurst/Oxford University Press.

Manor, J. (2007). In part, a myth: The BJP's organisational strength. In *Coalition politics and Hindu nationalism* (pp. 71–90). London: Routledge.

Manor, J. (2017). 'Ethnicity' in Indian politics, chapter eight. In J. Manor (Ed.), *Politics and state-society relations in India*. New Delhi/London/New York: Orient Black Swan/Hurst/Oxford University Press.

Mayer, A. (1960). *Caste and kinship in central India: A village and its region*. Berkeley and Los Angeles: University of California Press.

Srinivas, M. N. (2003). An obituary on caste as a system. *Economic and Political Weekly*, *38*(5), 455–459.

Sushmita (2014, January 11). Politics of massacres and resistance. *Economic and Political Weekly*, *49*(2), 41–45.

Varshney, A. (2002). *Ethnic conflict and civic life: Hindus and Muslims in India*. New Haven and London: Yale University Press.

Waghmore, S. (2013). *Civility against caste: Dalit politics and citizenship in western India*. New Delhi: Sage Publications India.

Waghmore, S. (2016). Challenging normalised exclusion: Humour and hopeful rationality in Dalit politics. In H. Gorringe, R. Jeffery, & S. Waghmore (Eds.), *From the margins to the mainstream: Institutionalising minorities in South Asia* (p. 153). New Delhi: Sage Publications.

2

ON THE VERACITY OF THE PROCESSES OF SOCIAL MOBILITY AND CHANGE IN INDIA

Manohar Yadav[1]

Introduction

The purpose of this chapter is to search for the realities underlying various mobility processes as perceived by scholars in sociology and social anthropology through some conceived frameworks whose logical contents remain doubtful and uncertain and, on that count, need verification. The mobility processes that occur in the Western world happen to effect real changes as in those situations assimilative propensities tend to be very strong and successful due to the absence of bias towards recognising the newly earned privileges of the historically subjugated groups; whereas in India, inbuilt prejudices and biases tend to devalue and degrade the achievements of the lower segments of society, leading to the loss of their integration or forced segregation. Under such circumstances, any process of mobility attempted through any given framework, at least from the viewpoint of the lower and backward castes, turns out to be insignificant and futile seen from its actual value as a potentially integrative force. In the Indian context, much has been said and hailed about the cohesive role of Sanskritisation in addition to the liberationist essence of the processes of modernisation and Westernisation. Hence, the nature of the consequences of the mobility processes of both of these frameworks (Sanskritisation and modernisation/Westernisation) has been examined here so as to ascertain their veracity with specific reference to some lower castes vis-à-vis the middle and the higher castes.

At the outset, one needs to adopt certain criteria to qualify the authenticity of mobility and change so that its effectiveness is genuinely felt and realised in actuality by the experiencing groups and individuals, and, in terms of those experiences, the success or the failure of a particular mobility process can be assessed and validated. The first criterion should be equality and assimilation, in that, those groups and individuals who are said to be undergoing processes of mobility should feel that their changed status position is valued and acknowledged as distinct from their earlier backgrounds, viz., caste and class, and, therefore, they are rated as equals in all respects by people and groups already belonging to upper caste/class backgrounds. It means a change in identity, occurring due to mobility, should receive identification in the wider context of caste and class; otherwise, it is likely to lose its practical worth and significance, thus making no real difference for the mobilising caste or class or the individuals concerned. And, of course, as a backing factor for this, another criterion should be to see whether the caste system becomes flexible, shedding its conservative characters seen in links with its operative order as a graded hierarchy as a whole. With particular reference to the idea of Sanskritisation, one can ask whether changes brought about by the conscious emulative acts of lower castes to be the upper castes lead to the changes in the old value pattern of caste positions, thereby ushering in a social revolution based on equality of treatment and mutual human regard. If not, what is the real nature of mobility and change in such a context? Can it be called superficial and just an apparent phenomenon?

With such questions in the backdrop, one can objectively examine theories of mobility and change with examples drawn from some corresponding ground realities.

Sanskritisation[2]

It is a well-known fact that Sanskritisation was popularised by one of the founding fathers and a doyen of Indian sociology, Professor M. N. Srinivas. According to him, through the process of emulating higher caste ways of life, the lower castes push themselves upwards in the scale of the caste hierarchy and bring about change in their status position. He termed such a course of action as "Sanskritisation". Srinivas lays down two specific conditions as necessary steps for a caste to be able to achieve mobility across the caste hierarchy through Sanskritisation. One is the adoption of the habits of vegetarianism and teetotalism, and another is the imitation of rituals, customs and other cultural practices of the dominant upper castes. A caste's adherence to this, for a generation or two, facilitates it to acquire a higher rank in the caste hierarchy (Srinivas, 1965, p. 30). Now let us look at its implications and practical possibilities.

Srinivas's assumption of mobility needs to be examined in terms of certain crucial characteristics of caste structure. First of all, caste is an endogamous unit and has a fixed position in relation to other endogamous groups placed in a static scale of hierarchy. So, endogamy and hierarchy form almost an inbuilt essence of caste. In fact, these two features dominate the entire mechanism of protection and persistence of the caste system in India and their continued existence puts a powerful ceiling on the ascribed and inherited status positions of the caste people. Hence, any mobility process which fails to neutralise these core elements of the caste system, does not really succeed in altering a caste's position. Perhaps out of his own realisation of this fact, Srinivas himself had expressed "doubts over the use of the term Sanskritisation" and even declared that "it (Sanskritisation) is an awkward term and be discarded quickly and without regret" (Parvathamma, 1970, p. 83). However, in teaching and academic circles, the concept of Sanskritisation still holds much respect notwithstanding the fact that it is also subjected to constant tests.

The assumption that the lower castes can be upgraded to the level of higher castes appears impractical as it faces powerful caveats in the process. The foremost question is: Can a caste be upgraded without degrading others? The report of the Karnataka Backward Classes Commission (Venkatswamy, 1985, p. 8) makes it clear that, "in Karnataka, each caste is so placed as to have some superior to it and some inferior to it and any caste at any rate cannot change its inferior or superior status". Scholars like D'Sousa (1967, pp. 192–211) opine that "it is not endogamy but the ranking of endogamous groups which is the essential mark of the caste system". He also prefers the term "hereditary groups" rather than "endogamous groups" for the units in the caste system (ibid.). Oliver Cox (1948, p. 11) writes, "each caste has its particular rank, defined by tradition and public opinion, and each one maintains it at all costs". Dr B. R. Ambedkar looks at the hierarchical arrangement of the caste system as something in which "a descending order of hatred and ascending order of reverence" prevailed. Hence, a few well-known sources treat "immobility" also as one of the definite traits of the caste system. According to the *Dictionary of Sociology* (Mitchell, 1968) "the caste system is rooted in the religious order and may be thought of as a hierarchy of hereditary endogamous groups with positions fixed and mobility barred by ritual distances between each caste". Even what D'Souza committed is incorrect when he says that "changes in the properties of caste bring about changes in its hierarchical relations" (ibid.). The caste's genuine properties being "endogamy" and corresponding "hierarchical rank" will not allow it to change its position even when it effects changes in its other properties like occupation, food habits, etc. In actuality, food habits and occupation cannot be the criteria for a caste to qualify for a superior or inferior position. Many respectable non-Brahmin castes have made non-vegetarianism a common food habit without losing the superior position traditionally accorded to them in a local caste hierarchy. In Karnataka, some of the dominant castes eat pork along with other red meats and yet they command a superior rank among castes which strictly avoid pork as being the most defiling and inferior of meats. So, the rank of a caste exists by itself with endogamy as its core and as a mechanism of maintenance as felt by both Dr B. R. Ambedkar and Edward Westermarck (Ambedkar, 1917; Westermarck, 1922). Thus, as long as endogamy and hierarchy continue to exist, there arises no question of upgradation or degradation of a caste as they act as strong cordons to the very process of mobility and its ensuing consequences like assimilation, integration and equality. The precondition, for a lower caste to be productively mobile, should either be an immediate higher caste moving down willingly to embrace a lower position of a mobilising caste or integrating it into its own rank, granting it the equality of status, no more or less than itself, in terms of a caste hierarchy; and through that process, there should also occur the merger of endogamous and rank boundaries of both the castes. Since this cannot happen, the status change deemed to have happened due to the emulation of upper caste ways of life, as asserted by the mobilising caste to have moved up from its earlier grade, should simply end up as an apparent and not a real truth.

Further, mobility through Sanskritisation in a caste context generates a sort of social friction. Lest its mobility endeavour be a superficial course of action, a mobilising caste will not be content with its mere emulative acts. So, while seeking upward mobility, it questions the superiority of upper castes, undermining all their credentials as the upper castes. Naturally, a kind of social revulsion flares up between the mobilising caste and the castes above it, which makes a case for a dismal consequence of Srinivas's Sanskritisation coin's one side. Turning to the other side of the coin, the outcome is even more depressing. The mobilising caste at the very same time condemns all other supposedly low castes as inferior and subordinate to it too. The belief here is, the more it hates, the more superiority it acquires over both the upper as well as the lower castes. In fact, it is an uncompromising and ruthless situation where everything remains virtually stagnant. Yet a mobilising caste being vehemently charged with the upper caste spirit is compelled to create a hostile environment between itself and the rest of the castes. And that is how the anti-social aspect of caste is again revitalised, strengthened and retained intact both in the mobilising castes and the rest of the castes, thus making no change in the old system of caste. Such mobility then does more harm to the castes than any real good.

Srinivas writes: "the Lingayats of Mysore claim equality with the Brahmins and more orthodox Lingayats do not accept food cooked or handled by Brahmins" (Srinivas, 1965, p. 30). This may be true. But equally true is the fact that, so far, the Brahmins have never given up their claim of superiority over the Lingayats. Also, it is not the conjoint experience of all castes in Karnataka to have sensed the tendency where Brahmins have accepted second rank importance to Lingayats and have thus incurred loss of status in a caste hierarchy. Even some of the untouchables do not accept food from Brahmins since they claim that in the beginning it was they who had been superior to all other castes. But such claims will not enhance the birth-ascribed status of the caste members and will not at all help their caste to upgrade itself in an all-time rigid framework of hierarchy. It is also true that emulation of the customs, traditions and food habits of a few noted castes by the rest of the castes turns out to be ridiculous and often a degrading exercise. To prove how Sanskritisation fails to bring about real mobility in a caste hierarchy, we may refer to Srinivas's own experience of goldsmiths. He says (1965, p. 30), "though they (goldsmiths) call themselves as Viswakarma Brahmins, wear sacred thread and have sanskritised their rituals, still then they are considered to belong to the left hand division of caste and no caste belonging to right hand division including the Holeyas (right hand untouchable caste) will eat food and drink water touched by them. Not only that, even their wedding procession was not allowed to go along the streets in which the right hand castes lived." If the process of Sanskritisation were to enable a caste to obtain a higher position, why it failed in the case of goldsmiths, is a question for which Srinivas had no answer. He simply states that "normally Sanskritisation enables a caste to obtain higher position in the hierarchy. But in the case of smiths it seems to have resulted only in drawing upon themselves the wrath of all other castes. The reasons for this are not known" (ibid.). In actuality, however, not only in the case of goldsmiths but also in the case of all other lower castes Sanskritisation fails, for it is based on imprecise reasoning. In some villages of Belgaum district, the members of lower castes like Kuruba, Beda, Kumbar, Madivala and Chaluvadi claim the status of Lingayats since they wear Lingam (symbol of phallus representing Lord Shiva), apply Vibhuti (sacred ash) on the forehead and perform Shivbhajans (singing devotional songs in praise of Lord Shiva) in the evenings. Also, as part of that assertion, they have strictly abstained from eating non-vegetarian food and have become teetotalers too. These castes have been followers of Veerasaivism (Lingayatism) ever since its birth in the twelfth century AD. However, there has not occurred any change in their ascribed status. Instead of making equality common to each of them, they all accept the relative superiority or inferiority of themselves and of other different castes in a local caste hierarchy. In this context, and as noted earlier, one more potent issue comes to light. If non-vegetarianism and teetotalism were the governing principles of low gradation of castes in a hierarchy, why have they failed to degrade those Brahmins and other upper castes indulging in the consumption of intoxicating drinks and all kinds of dishes for generations together? After coming in contact with the British, and of course other foreign people, many upper caste communities have joined them in eating non-vegetarian food including beef and consume alcoholic drinks too.[3] But, by doing so, never have they experienced a fallen state of status than one accorded to them by the caste tradition. In fact, these castes claim the most enhanced position and continue to be even more conservative in matters related to caste. It is also true that teetotalism is more of a personal trait than an attribute of an entire caste community. Furthermore, can any other caste be converted into a Brahmin caste just because its members have learnt to wear the sacred thread or imitate

Brahmin ways of life? At least at present, it is not likely to happen, and, when it is not likely to happen, it has no meaning.[4]

Further, castes, being self-sufficient cultural units, need not imitate any other castes. Cox (1948, p. 37) admits that "each caste has its own God given dharma, its religious way of life and natural priesthood. Indeed, we may think of caste dharma as Hinduism in microcosm". According to Dr Ambedkar:

> a complex congeries of creeds and doctrines is Hinduism. It shelters within its portals monotheists, polytheists and pantheists; worshipers of the great Gods Shiva and Vishnu or of their female counterparts, as well as worshipers of divine mothers or the spirits of trees, rocks and streams and the tutelary village deities; persons who propitiate their deity by all manner of bloody sacrifices, and persons who will not only kill no living creatures but who must not use the word 'cut'; those whose ritual consists mainly of prayers and hymns, and those who indulge in unspeakable orgies in the name of religion; and a host of more or less heterodox sectaries, many of whom deny the supremacy of the Brahmins, or at least have non-Brahmin religious leaders.
>
> (Moon, 1987, p. 15)

Because of this chaos, castes do not have a blending spirit in them. Hinduism, in its own unique way, recruits castes, but never integrates them into homogenous units. Instead, it grants them social ranks with specific positions and fixes them in a scale of hierarchy with an injected degree of disdain or veneration for castes below or above each caste respectively.[5] Indeed, Hinduism lacks a universal essence, which can bind its followers into a single whole. Under such circumstances, amalgamation or assimilation is almost impossible. Given all these hard constraints, mobility through Sanskritisation tends to be more like a mirage chasing exercise.

Modernisation/Westernisation

Let us now turn towards the quandaries we run into as we go about exploring the likely nature of the social mobility deemed to have occurred amongst the lower castes in the context of much hailed modernisation. Here, as well, the views of a few key scholars can be put to critical examination. McKim Marriott, one of the early scholars known to have worked a great deal on Indian society, felt that, with the rise of modernisation trends, there transpired a circumstance of social disintegration moulding the system on the lines of commercial relations of status. He writes, "there is a disintegration of social system based on group or corporate relations", and, the reasons he gives include "the double processes of Sanskritisation and Westernisation"

(Marriott, 1967, p. VIII). As we have already seen, disintegration of the social system may not take place through Sanskritisation as in its case, and, as emphasised by Srinivas himself, mobility works only at the level of the caste group concerned and not at the general level of individuals spanning the entire caste system. In fact, Sanskritisation strengthens group solidarity, by increasing integration strictly among its members who desperately try to be graded as an unequal but as a high-level social status group in a given time and place of caste hierarchy. However, disintegration is a potential possibility under the process of Westernisation in that being itself a modernisation process, Westernisation leads to a decline in occupational specialisation affecting the interdependence of castes. But this view needs to be accepted with some caution because, whatever its form, Westernisation happens to be a mere trimmed garment of tradition in India, and, in spite of effecting changes in castes' occupations, utterly fails to bring about disintegration of the social system based on caste. The Karnataka Backward Classes Commission, headed by late L. G. Havanur, wrote in its report (1975, p. 90):

> We tried to avoid caste, but we found it difficult to ignore caste in the prevailing conditions. In modern times, any body can take to any profession. The Brahmin taking to tailoring does not become tailor by caste. Nowhere is his social status lowered as a Brahmin. A Brahmin may be seller of boots and shoes and yet his social status is not lowered thereby. Social backwardness is, therefore, not today due to the particular profession of a person but due to caste itself.

Thus, changes in occupational specialisation and the interdependence of castes do not disintegrate the social system of caste because of its enduring elements – endogamy and hierarchy – as they together grant and restrict birth ascribed status to members belonging to different castes. Also, occupational change, since it questions the credentials of caste related occupational positions, on occasions creates humiliating experiences of a sort. The Havanur Report further says:

> Even a high court judge of a low caste is subjected to all sorts of insinuation and humiliation, by higher caste Hindus, as his uneducated village brother is subjected by the higher caste Hindu villagers. (ibid.)

Hence, for social mobility to be truly factual, it needs the neutralisation of caste related status positions, perhaps top castes to start with, as they happen to be the reference points. However, that appears a very remote possibility. Caste demonstrates to be equally important both in the context created by modernisation and the persisting traditional situation of society. Almost like a shadow, a person must carry his caste, wherever he goes, be it school, college, politics, administration or any other domain. This tendency of caste was acknowledged long back by S. V. Ketkar.[6] He (1909) writes, "if Hindus migrate elsewhere in the world, caste system would become world's problem". And, one can witness many caste associations have sprung up in USA, UK, etc. as the migrant Hindus have opted for such organisations. So, though people profess to be caste neutral, they develop their social and blood bonds through marriage within their own respective castes. Members of different organisations, whether it is a political party, trade union or club, enter into caste-based conspiracies, suspicion, and distrust of each other. People may project an outward appearance of being free from caste considerations, yet in practice, they do everything as adherents of their caste concerns. Hence Westernisation, urbanisation, or to use a more specific blanket term "modernisation", has not helped India escape from its caste noose. Anil Bhat (1975, p. 307) writes: "the gap between the higher status castes and the lower status castes is greater in urban than in rural areas". Urban areas are known for their cultural, educational and social activities. Also, they run mass communication channels like TV, radio, cinema and drama. More importantly, they happen to be industrial hubs and commercial centres as well. These ingredients, indeed, add glamour to the urban establishments. Nevertheless, they have also become equally casteist bastions, because they all ultimately depend upon caste for their growth and development. In the entire commercial streets of Bangalore or Bombay, hardly any member of a lower caste can be found to be owning a shop or a business enterprise. Obviously, whatever change is deemed to have come about due to Westernisation seems to be just apparent and not real. As if by its reinforcing course of diligence, the entire caste system reproduces itself in the urban context. And, this may be the reason why the Dalits, the Backward Castes and Tribes outnumber any other community of people in urban slum settings. Hence, there is no question of Westernisation disintegrating the social system based on caste or corporate relations.

Further, change in occupational specialisation as advocated by McKim Marriott is all right. But that is not so important. What is important is what sort of change has occurred in occupational specialisation. A caste Hindu claiming to be liberal because of Westernisation will not be ready to take up such occupations as scavenging, street sweeping, shoe-making or carrying dead animals. He will not dare do that since he can take up only such occupations which will not only not harm his caste prestige and status but also help him enhance the same to still greater heights. Only with such specification will he do away with his traditional occupation. Hence, in most cities, scavenging is still an exclusive job of the lowest castes. Not only that, no top caste person can be found engaging in the activities of stone or wood cutting, mud carrying, etc. on the sites of construction of roads, buildings, bridges, dams etc. Therefore, a change in occupational position has no meaning insofar as it does not impinge on one's ascribed caste status, or corresponding stigma and prestige.

In a similar vein, D. Mandelbaum (1966, p. 1538), also opines that "the lowest and the poorest have got new opportunities, (due to political changes after Independence), to change their self-identification". But this wishful thinking too suffers from a fundamental flaw. The key question is, however willing a low caste is to change its identification, will others recognise that identification? That it is unlikely to happen can certainly be upheld as a matter of stark reality. It is true that due to changes in political conditions, lower castes have entered into politics, and, of course, professions like law, medicine and other public services (no matter what percentage they actually form in these fields). Some of them have also given up thatched huts and have started living under tiled roofs. A few of them have even stopped their children from going to work as wage labourers, street sweepers or as scavengers, and, have pressed them to attend schools. No doubt, this is a striking change. But it is only one side of the picture. On the other side, what we see is higher castes living in buildings with ultra modern amenities, making a big dent in commerce, trade and industrial ventures, occupying crucial positions in public domains and moving in cars and aeroplanes. In all the important commercial streets of India, at least up till now, hardly could an untouchable be seen to have made any space for himself. So, virtually, both the caste Hindus and the lower castes almost remain in the same old mode of socio-economic inequalities with cultural crevices prevailing between them. More specifically, it is remarkably true of the untouchable castes of modern India. But then it substantiates the fact that the modernisation process has only strengthened the base of various castes without actually affecting their hierarchical differences and corresponding degree of benefits. In an era where processes like globalisation turn out to be more exclusivist in disposition, caste bonds develop instil unusual ways, and opportunities of affluent life tend to proceed along caste lines. Eminent sociologists like N. Jayaram and Satish Saberwal also note that there is a "strong tendency for access to opportunity – and information and insight – to flow through close, familial and *jati*-like relationships; this has consequences for entering educational institutions and, subsequently, the world of work" (Saberwal, 1996; Jayaram, 1987).

In the political domain, specifically, a sort of distinctive untouchability is experienced since many of the highest positions bestowed on untouchables have only proved to be illusory or useless for them. Being an untouchable, one of the working secretaries of the district unit of a dominant national political party in Karnataka opines as follows:

> I don't have a voice in the party as it is controlled and owned by upper caste bosses. If I raise my voice, and try to be smart, I will be thrown out of the party. And in my place, some other untouchable

> who listens to the party bosses (actually means upper caste bosses) will be brought. I am a general secretary in the party since several years. But never have I been able to protect either my own individual interest or the interest of my community. Being a lawyer, I can articulate, but for no use. On several occasions, I felt humiliated while seeking a party ticket for myself, or for someone who is worthy like me, whereas Hamals (luggage carriers in a bus stand) belonging to my community got tickets since upper caste party bosses tend to opt only for such type of gullible people.

This trend has been in practice ever since Independence. While blaming the political system as a whole, B. P. Maurya, an eminent Ambedkarite and nationally known political leader, once said as follows:

> This system does the Scheduled Castes no good because the people in the reserved seats belong to the party in power and are often incapable persons. Although they are educated, they do not speak out against the party in power. They do not represent their people to the party and government but represent the party in power to their people.
>
> (Isaacs, 1965, p. 185)

Like untouchable politicians, the untouchable administrators also suffer the same fate, because colleagues around them belong to higher Hindu castes. Some of the government servants belonging to the untouchable community still feel as follows:

> The non-cooperation in some of the offices, due to untouchability, is so horrible that all other caste men and women join on one side while systematically isolating untouchables. They unitedly conspire, and do all possible harms to the untouchable coworker, and sometimes till she or he is effectively punished. Such an experience is a sort of shadow boxing. The untouchable person concerned feels the pain of the blow but does not understand the source of it. And there is no end to such kind of sufferings.
>
> (Yadav, 1989, p. 18)

Havanur (1975, p. 276) views such kinds of tragedies as follows:

> Adult franchise and spread of education coupled with discriminatory protective measures in favour of certain castes have, no doubt, given scope for lower caste individuals to enter into political fields, democratic institutions and the new professions. It is to be noted here that the lower castes in politics, in democratic institutions and

new professions are ineffective, as their higher caste colleagues look down upon them as low and inferior.

Thus, even now, caste operates as a widely instilled social institution. Whether in politics, administration or any other field, it keeps raising its ugly head. But its thorough saturation forces the lower castes to keep their identity intact and ever fixed so as to be of no value. Under such circumstances, an untouchable's asserted new class identity, achieved due to democratically available educational, administrative and political means, fails to materialise, as it cannot really change his status of being born as an untouchable. Ultimately, however determined an untouchable is to change his identification, the fact that others do not recognise the change in literal terms indicates that his asserted identity is apparent and not real and hence fails to bring him real benefits of change as discussed by Mandelbaum.

Owen Lynch (1969, p. 204) subscribes to the idea that caste uses itself to get "scarce goods such as power, prestige, wealth and education in the new socio-political environment brought about by the independence of India". According to him, this has become possible because of the caste's "adaptive structure". He also says that the adaptive characteristics of caste are of two types: conservative and creative. In its conservative aspect of adaptation, caste does not allow itself to disintegrate, as it remains safe, sound and stable. However, while keeping itself unharmed, it mobilises its members to get the above stated "scarce goods", which actually are caste external. Hence, an attempt on the part of caste to get these things is, to Lynch, "a creative aspect of caste".

Of course, what Lynch maintains is precisely true. But when it comes to the question of analysis of the real utility of the caste's adaptive quality, his argument doesn't offer a much convincing answer. One can ask whether the conservative aspect of caste is quite a new one to it. That the conservative character of caste is as old as caste itself is beyond one's doubt. Because by its very genesis, caste principally stands out to be definite and rigid both in nature and deed. So that aspect of it cannot be new to it. And, because of that peculiarity, not only has caste strongly adapted itself to the new socio-political environment created after Independence, but has also been surviving shocks of many more such historical transformations over time. Exceptionally, in all these times, caste has successfully shielded its central essences like endogamy, hierarchy and corresponding sanctions and dues. Hence, the conservative spirit of caste is bound to continue so long as these core ingredients of it continue to persist. Further, the conservative and creative aspects of caste are functionally related. Caste continues to be conservative only because it wants to be creative. Thus, the creative aspect of caste is also not new to it. The creative and conservative aspects together preserve and protract the whole structure of the caste system in India. However, after Independence, so as to make the most of new opportunities, caste has become more dynamic than ever before and has added fresh vigour to both its conservative and creative facets. This means, while adapting to a new environment, for its absolute advantage, caste has become excessively active. In addition, to keep in tune with the spirit of the time, it has often been changing its forms and spirit as well. Conceivably, there is a sort of compulsive reason for a caste to behave so. Why do castes become adaptive? It is only to become more intact and rigid no matter how often the very forms of such rigidity change. Caste accepts liberal ideas of the West to the extent such gestures help it preserve anti-liberal concerns. A high caste man may speak fine English, prefer to be dressed in a suit, etc., yet he equally perseveres to be conservative in his outlook and values. This attitude and its corresponding consequences as generated by the entire interactive membership of the castes, in the unusual context, do give a new lease of life to many of its old functions. If an untouchable caste, as asserted by Lynch, functions as an association or as an interest group, and mobilises its members for political ends or action, so do all other castes also for the very same purpose. While castes mobilise for achieving their ends, they, at the same time, actually try to do it at the cost of other castes. Such an act on the part of caste reflects its old characteristic, taking on a new form. Yet it has in it a process of exploitation, discrimination and injustice or all that one can consider as the basics of the caste system. There is a severe competition among castes to get the ever-growing new ends fulfilled. And in this competition, it is the higher castes, because of their traditional privileges and power, which stand to gain more and win against the lower castes, thus keeping the lower castes in the same low state. When castes' adaptation to a new environment fails to produce the desired results, in terms of impartiality and egalitarianism, any change brought about by such a course would prove to be mostly futile and ineffective for the lower castes. So such a change from their viewpoint becomes a mere surface look or a peripheral change without any inner force in it.

Finally, Lynch makes an important case for a caste in the new context of India's post-Independence. He asserts that in the new environment, caste is transforming itself into a secular entity. But, so far, at least as most of the lower castes' experience shows, India is yet to evolve the necessary preconditions for the successful working of democracy. In the absence of such preconditions, one could be doubtful about the possibility of caste becoming a secular entity. The foremost question is: how can a caste become secular when it uses itself for achieving its own exclusive communal interests with no tolerance for other castes? B. P. Gajendragadkar (1965, p. 60) writes: "The spirit of toleration is the foundation of the theory of secularism." When the entire social system of caste is imbued with an anti-secular spirit, how can a caste unit become secular? According to Dr Ambedkar, "Caste basically stands for contempt and hatred". This is clearly obvious in the way castes contest to retain the power base under their utter control.[7] While suggesting the necessary preconditions for the successful working of democracy, Dr Ambedkar felt, "In a society where classes clash and are charged with anti-social feelings and spirit of aggressiveness, the government can hardly discharge its task of governing with justice and fair play. In such a society, government, even though it may, in form, be a government of the people and by the people, it can never be a government for the people. It will be a government by a class and for a class" (Moon, 1987). What Dr Ambedkar (ibid.) identifies is "lack of reciprocity of interests among groups and individuals" often tending to produce anti-democratic and anti-secular consequences in the exceptionally unique caste context of India. Even M. N. Roy opined: "India cannot become a secular state, because the preconditions necessary for it are quite absent." In light of the above realities, the following conclusions can be drawn.

First, mobility achieved through the process of Sanskritisation tends to generate mutual rivalry between higher and lower castes rather than promoting progressive advancement in the caste hierarchy. In the absolute absence of possibilities for the absorption of each other's membership into the common caste-bond, the castes in the context get involved in claims and counter claims for superiority over each other. In such a situation, it is necessary that the higher caste should step down from its hierarchical position to elevate the lower one in its place, which is unworkable because of the caste's strong obsession with its status position. So, the mobility deemed to have occurred in the status position of Sanskritising caste tends to be illusory rather than real. The precondition for mobility, through Sanskritisation, to be truly factual, is that there should be an opportunity for members of lower castes to share membership with the upper castes or for members of upper castes to embrace the lower ones. This is not possible because endogamy caps such opportunities.

Second, changes in the specific occupations of castes have not disintegrated the social system based on corporate relations. A Brahmin does not become a Chamar by caste merely because he takes to the occupation of shoe-making. Similarly, the low caste men may achieve coveted positions in the government or other echelons of the power structure and identify themselves as higher classes, but in the absence of equal identification by similar classes belonging to upper castes, they remain suspended as parallel classes in their respective caste units.

Third, the adaptive element of caste keeps on adjusting with changes brought about by greater processes like Westernisation and modernisation without harming its inner features like endogamy, hierarchy, non-commensality, or other marks of distinction, thereby compelling caste and its corresponding membership to remain intact with the status quo of either superiority or inferiority, thus making change meaningless and void. The worst victims of this type of systemic behaviour have been the untouchables, because their achievements in terms of gaining power, prestige, wealth and education have proved to be valueless and worthless.

Finally, there are two issues which continue to underpin any analysis of Dalits and the other lower castes' situation in India. One is historical and structural in nature, something that has an inherent tendency to reinforce itself so that it continues to preserve itself and persist further in one form or the other. That is how a whole lot of old inequalities are reproduced in new forms. And another issue is: how hard it is for any emerging system like democracy or modernity in India to overcome the continuing influence and control of what is said to be historical and structural in nature. Here, a sort of war has been going on between historical structures of inequality and liberal democratic ideals of equality. And in this war is locked the supposed emancipation of the so-called lower castes or the Dalits in India. At least, so far, the triumph has always been on the side of enduring structures of historical inequality. This is not to say that nothing is happening in this country insofar as the oppressed castes and communities are concerned. But only so much is happening which ultimately gives an impression that nothing is happening. It is something like what sociologist Durkheim termed as "motion without progress". Indeed, a very confusing and puzzling situation.

Notes

1 This is an expanded and revised version of a commentary published by the author in *Economic and Political Weekly* titled "Mobility through sanskritisation: An apparent phenomenon?", *51*(24), 12–14 in 2016.
2 Some arguments, because of their continuing relevance, have been derived from my PhD thesis entitled "The politics of socio-economic inequality: The riddle of untouchables in Belgaum District" submitted to Karnataka University, Dharwad, for award of PhD degree in 1988.
3 In states like Kerala, meat consumption is very common and for non-vegetarians, eating beef is also not a taboo. Alcoholic drinks are widespread and most common too.
4 M. N. Srinivas seems to be silent on the issue of why vegetarianism is superior to non-vegetarianism and therefore worthy of emulation.
5 Srinivas called the British the neo-Kshatriyas.
6 Ambedkar regarded Ketkar as one of the original thinkers on caste.
7 That is how regionally powerful castes in India by the sheer mustering of their respective caste resources have succeeded in holding sway over political power. In Karnataka, Lingayats and Vokkaligas dominate the political scenario while in Maharashtra, Marathas have sway over it. In Karnataka, untouchables, being the single largest community, have not been able to produce one chief minister to date, whereas even minute upper caste communities have been able to do so to their credit at least once in a while.

References

Ambedkar, B. R. (1917). Castes in India: Their mechanism, genesis and development. *Indian Antiquary, XLI*.
Bhat, A. (1975). *Caste, class and politics*. New Delhi: Manohar Publications.

Cox, O. C. (1948). *Caste, class, & race: A study in social dynamics*. New York and London: Modern Reader Paperbacks.

D' Sousa, V. S. (1967). Caste and class. A reinterpretation. *Journal of Asian and African Studies*, 2(3–4).

Gajendragadkar, B. P. (1965). *Law, liberty and social justice*. Bombay: Asia Publications.

Havanur, L. G. (1975). *Report of the Karnataka Backward Classes Commission*. Bengaluru: Karnataka Government Publications.

Isaacs, H. R. (1965). *India's ex-untouchables*. Bombay: Asia Publishing House.

Jayaram, N. (1987). *Higher education and status retention: Students in a metropolis*. New Delhi: Mittal Publications.

Ketkar, S. V. (1909). *The history of caste in India: Evidence of the laws of Manu on the social conditions in India during the third century AD: Interpreted and examined: With an appendix on radical defects of ethnology*. Ithaca: Taylor & Carpenter.

Lynch, O. (1969). *The politics of untouchability*. Agra, India: Columbia University Press.

Mandelbaum, D. (1966, December). Review of the book India's ex-untouchables by Harold R. Isaacs. *American Anthropologist*, *68*(6), 1536–1538.

Marriott, M. (1967). *Village India*. Chicago: Chicago University Press.

Mitchell, G. D. (1968). *Dictionary of sociology*. London: Routledge and Kegan Paul.

Moon, V. (1987). *Babasaheb Ambedkar: Writings and speeches* (Vol. IV). Government of Maharashtra Publications.

Parvathamma, C. (1970). *Politics and religion* (p. 16). New Delhi: Sterling Publishers, (P) Ltd.

Saberwal, S. (1996). *Roots of crisis: Interpreting contemporary Indian society*. New Delhi: Sage Publications.

Srinivas, M. N. (1965). *Caste in modern India and other essays*. Bombay: Asia Publishing House.

Venkatswamy, T. V. (1985). *Report of the Karnataka Backward Classes Commission*. Bengaluru: Karnataka Government Publications.

Westermarck, E. (1922). *The history of human marriage*. London: Macmillan and Co Ltd.

Yadav, M. (1989). *The politics of socio-economic inequality: The riddle of untouchables in Belgaum district* (Unpublished PhD thesis). Karnataka University, Dharwad for award of PhD degree in Anthropology.

3

RECASTING POLITICAL MOBILITY

Scheduled Caste Gram Panchayat presidents and the "dominant lower castes" in Tamil Nadu

Anand Inbanathan

The 73rd Constitutional Amendment Act, passed by parliament in 1993, brought into effect a clear statement of the importance of the Panchayati Raj institutions in the Indian political structure. While Panchayati Raj existed even before this Constitutional Amendment, it did not have a significant place in the Indian Constitution, and in their functioning, the Panchayats depended too much on the political will of state governments to ensure that Panchayats were supported, or even held elections at regular intervals. States that had not held regular elections to the Panchayats now had to change, and the 73rd Constitutional Amendment made it necessary for the states to hold Panchayat elections regularly, and so far as we know, that has actually been implemented. This enactment has also ensured that there is a significant number and proportion of seats in the Panchayats reserved for various groups, for those of the Scheduled Castes, and for women.

While Panchayat seats have been occupied by people of these groups (i.e. Scheduled Castes, and women, as the case may be) in the reserved seats, there is another issue that needs to be considered. This is whether the elected representatives are able to freely function as elected representatives, and in some cases as presidents, or whether there are constraints and barriers that restrict or even prevent them from carrying out their political duties, as Panchayat presidents or representatives. This chapter considers this question in the context of the Panchayats that have presidents who are from the Scheduled Castes, and how they are able to function in a state which has shown a clear disinclination to allow presidents of the Scheduled Castes to carry out their duties as presidents.

Over the years, even with the reservation of seats provided for members of Scheduled Castes, and among the executive positions such as president and vice president of the Panchayat (the vice president's post is not reserved for Scheduled Caste members in all states), the level of participation of those from these reserved groups has not been of a high order. It has been seen across the country that they are mostly proxies, taking instructions from dominant individuals of the villages (Mathew & Baviskar, 2009). In Uttar Pradesh, Leiten and Srivastava (1999) wrote that their Jat respondents stated that it would be an "inversion of the natural social order" if members of the Scheduled Castes became Gram Panchayat *pradhans*. Local people whom Leiten (1996) had talked to, had suggested that having Scheduled Caste individuals in the panchayats as ordinary representatives was acceptable, but not as pradhans. He also stated that representatives of Scheduled Castes were "namesake" or nominal members, and had no real power or authority to be involved in decision making in the Panchayats. There was also the feeling that a Gram Panchayat pradhan will, in a sense, be their overlord, i.e. the local castes thought so, and they found it completely intolerable to have a Scheduled Caste dominating them. It is another matter that in an objective sense, Gram Panchayats are not institutions of any high degree of power or authority, and the president/pradhan cannot wield an extraordinary power as Gram Panchayat pradhan. Even in states with a higher proportion of Scheduled Caste members, and therefore, a more assertive number of Scheduled Caste people, there are clear indications that locally dominant castes can and do control the functioning of Panchayats which may have significant proportions of Scheduled Caste members, and also if a Gram Panchayat Sarpanch's post was reserved for a member of a Scheduled Caste (Martin, 2015).

Whatever be the factors that impede the functioning of presidents who are from Scheduled Castes, it has been well demonstrated in various states that they are not usually able to function effectively or even in a reasonably independent manner as presidents of Gram Panchayats. In Tamil Nadu, the situation has taken a grim turn on several occasions in the past 25 years (i.e. since 1993). In one case, a Scheduled Caste president of a Gram Panchayat in Madurai district was killed because he had ignored the warning of a local dominant caste and had filed his nomination papers to become president, against the explicit directions of this caste group. A few months after he became Panchayat president in Melavalavu, Madurai district, he was killed in broad daylight, along with others, numbering six Scheduled Caste men, in June 1997. The identified and later convicted men were of the Kallar caste, a dominant caste in that area. There have been more news reports of presidents who belonged to Scheduled Castes, and from other districts of Tamil Nadu, who had been killed in the years after this occasion of the murder in Melavalavu, and the guilty men were later identified as members of locally dominant castes (Mathew, 2003; Kannan, 2007; Frontline, 2013).

What we will discuss here is the functioning of presidents of Gram Panchayats who are from Scheduled Castes. This includes both men and women who have been presidents. In ordinary circumstances, one may presume that on becoming presidents of the Panchayats, the new incumbents would be able to access a new status, and thereby also the authority and power that are vested in this position. That, however, has not happened, and although the reservation of seats has brought members of Scheduled Castes into the Panchayats as representatives, and as presidents, that has not automatically conferred any authority or power on these representatives. Thus, it is also not immediately seen that there is any degree of political mobility of these groups by virtue of their occupying positions which are ostensibly of some importance, and also carry authority, but such authority is not reflected on the people of Scheduled Caste groups. This chapter draws from the field study conducted in six Gram Panchayats in Tamil Nadu.

Castes and dominant castes in Tamil Nadu

The Scheduled Caste population in Tamil Nadu, according to the 2011 census, is just under 20 per cent. They are not uniformly distributed in all the districts, and while Kanyakumari has a Scheduled Caste population of just 4 per cent, Thiruvarur district has nearly one-third of its population comprising people of the Scheduled Castes (32.4 per cent, 2001 Census).

In his conceptualisation of castes and their positions in society, M. N. Srinivas suggested the concept of "dominant caste" to designate a caste which had a preponderance in numbers, wealth, and political power that together can place them in a dominant position vis-à-vis the other caste groups that live in the village (Srinivas, 1959). The usual presumption is that these castes have a prime position, or are superordinate to other castes in the village. That does not by itself indicate that the castes have a superior position across several villages, or in a wider area of the state (Mayer, 1958). The dominant castes have a place in the village that gives them some control over other castes of the village, and the functioning of the village also owes much to the elite of these castes, who are able to manage the functioning of the village. There are villages which have men, who are referred to as "big man" of the village, in Karnataka, who have a position in the village that enables these men to control the Gram Panchayat and other activities in the village (Inbanathan & Gopalappa, 2003; Chandrasekhar, 2011). As we describe below in greater detail, Tamil Nadu too has villages which have dominant castes, as well as men who function as a "big man" of the village.

An unusual occurrence in Tamil Nadu is not just that there are dominant castes, but they are not in the upper half of the caste hierarchy. They are mostly in the lower half of the caste hierarchy, and just above that of the Scheduled Castes. Arun (2007) calls these castes as "dominant lower castes".

The caste that is dominant in each local area has been identified, and we have tried to discern what they do and how the Gram Panchayat functions in the area of their dominance, with a Scheduled Caste member as Panchayat president. The preponderance of dominant castes in Tamil Nadu is usually over smaller areas, and thus, the dominant castes in different districts vary (Lakshman, 2011; Karthikeyan, Rajangam, & Gorringe, 2012). There could be several factors that influence the position and functioning of Scheduled Castes in the Panchayats of the state. Among them is their numerical strength in any particular area. If Scheduled Castes are numerically strong in a village, they could be less dominated. On the other hand, if they comprise only a small proportion of the village population, then they are under the dominance of local and higher castes who are found in the village. The dominant castes may be designated as Backward Castes, rather than being included among high or upper castes. Scheduled Caste members are usually employed by landowners from these castes, and are dependent on them for their livelihood. If these persons were to be elected representatives and presidents, their ability to function independently, or even function at all as president, is severely circumscribed.

Caste consciousness in Tamil Nadu is of an extremely high degree. Locally dominant castes can take exception to any event or act that they consider as "improper", and not in keeping with the low status of Scheduled Castes. When they do disapprove of any action of the Scheduled Castes, these upper castes can and do take severe and even violent action against them. A further reason for the antagonistic stance of castes higher than Scheduled Castes is the affirmative action taken by the state, where seats are reserved for the Scheduled Castes, which members of the upper castes felt was at their cost, and they were deprived of what was due to them (Vincentnathan, 1996; Radhakrishnan, 2002; Kannan, 2007).

In states such as Karnataka, a constituency is reserved for any group for a term of five years, and then at the next election, the reservations are moved to another group, or they are made "unreserved" constituencies. In Tamil Nadu, the same constituency is reserved for the same group for a period of two terms, i.e. in two successive elections the constituencies are reserved for the same groups. This has aggravated inter-caste relations to the detriment of the Scheduled Castes. It is another matter that even when Scheduled Caste members are elected to the post of president, various means have been employed to get around having a Scheduled Caste as the *de facto* functioning president, and they are for the most part nominal presidents. A vice president of an upper caste would be the person managing the activities of the Panchayat, and would function as the *de facto* president. We have examined the manner in which Scheduled Caste presidents have functioned in six Gram Panchayats in different districts of Tamil Nadu, which also gives an idea of whether they can or do function as independent presidents. These Panchayats have different castes who are dominant in these villages and the Panchayats.

Gram Panchayats and Scheduled Caste presidents

Six Gram Panchayats were selected in all, with the main criterion that all these Panchayats had presidents who were of the Scheduled Castes. Tamil Nadu has Gram Panchayats of various sizes, in keeping with their population in the village, in comparison to states where every Gram Panchayat has approximately the same population size (for example in Karnataka).

Two Gram Panchayats were selected in Cuddalore district, viz. Pallipattu Gram Panchayat, and Annaveli Gram Panchayat, which have Vanniyars as the dominant caste in their areas. Vanniyars of the villages comprising the two Gram Panchayats are also the predominant landowners. Members of Scheduled Castes in these villages depend on the Vanniyars for their livelihood and income. The functioning of both the Gram Panchayats are under the control of Vanniyars, and it is the Vanniyars who take all the decisions of the Panchayats. A Vanniyar man who is also a Union Panchayat councilor dominates the Pallipattu Gram Panchayat. The vice president is a Vanniyar who is associated with the Vanniyar Union Panchayat councilor, and together they manage the Panchayat. The president cannot do anything without their approval. Local residents observed that while the two men who managed the Panchayat stated that they were engaged in development works for the benefit of the Gram Panchayat residents, a substantial part of the funds went to the Union Panchayat member and the vice president of the Gram Panchayat, and much of the "development" work was not implemented, which was clearly evident (i.e. lack of development). The president was also kept under the control of the Vanniyars through the support of the police sub-inspector of the area, a Vanniyar himself. The Panchayat secretary too functions on the instructions of the Vanniyar vice president, and not the Scheduled Caste president. The population of Vanniyars and Scheduled Castes in this Panchayat area is about the same, but Vanniyars are the dominant group, and not the Scheduled Castes.

Annaveli Gram Panchayat is the second Panchayat in Cuddalore district. It had a woman president (the president's post had been reserved for a Scheduled Caste woman). The Vanniyars are larger in number than the Scheduled Castes in this area, and Vanniyars are predominantly the landowners of the area, while Scheduled Castes are largely landless. The reservation of the Panchayat president's post was strongly opposed by the Vanniyars, by harassing the Scheduled Caste people, though ultimately they had to give in, as the state government authorities had notified this Panchayat as reserved for a Scheduled Caste woman. The vice president, who is a woman of the Vanniyar caste, manages the Gram Panchayat with the help of her husband who was the previous Gram Panchayat president. He was as of this time the Union Panchayat president. The cheque book for operating the Gram Panchayat account was kept in the vice president's house, and it was the president who had to go to her house and get a cheque leaf. This particular woman of a Scheduled Caste was elected as president with the support of the Vanniyars, with the view that she could be easily dominated and controlled.

In this Gram Panchayat, Scheduled Caste representatives were not allowed to speak. Very few development activities were carried out in the Scheduled Caste colony. A Scheduled Caste woman being the president of the Gram Panchayat has not made any difference to the people of the Scheduled Castes in the Gram Panchayat, since in any case she does not take any of the decisions of the Panchayat.

Theni district has the two Gram Panchayats Sinthalaseri Gram Panchayat and Endupuli Gram Panchayat. The Udaiyar caste is the dominant caste in the Sinthalaseri Gram Panchayat. Two Panchayat presidents (i.e. from earlier terms) were from this dominant caste group, and the Udaiyars were not happy with the situation where the post of Panchayat president was now reserved for a person of the Scheduled Castes. While there is no single individual who appears to be a dominant personality here, the caste has a dominant position in the area. The vice president is from this caste. The Panchayat president is respected less than the Panchayat clerk who is of a higher caste. Even schemes for the Scheduled Castes were not implemented properly so that they would benefit, including housing schemes. They were not treated with any degree of priority, and the president who had to forward these requests to the block level Panchayat administration was forced to wait for a long time because he is of the Scheduled Caste, and the block level administration was reluctant to act on the recommendation of a person of a Scheduled Caste.

The second selected Gram Panchayat in this district, Endupuli Gram Panchayat, is dominated by the Thevars. Pallars and Paraiyars are the largest groups of Scheduled Castes, and the smallest group of Scheduled Castes is the Arunthathiyars. One Thevar family has been a dominant group in the Panchayat, where one man, who was the head of his family, was district secretary of the DMK party. He became the first Gram Panchayat president of this Panchayat. After his term was over, his nephew became the next Gram Panchayat president, and then it was the turn of the head's son as president. The family was not pleased with the way the new reservation policy had played out, with a Scheduled Caste woman who became the Panchayat president. In any case, she was chosen by this Thevar family, and is of the Pallar caste, and considered as someone they could easily control. She is not even allowed to sit on the chair of the Panchayat president. She does not know anything about what to do in the Panchayat, and is advised by her husband. But, ultimately, it is the Thevars who run the Panchayat. The vice president is a Thevar, and he tells her what to do. The president only has to sign documents, whenever she is told to do so, and does nothing else.

Among the reasons for choosing a Pallar as president of this Panchayat is that Pallars and Paraiyars, two groups of Scheduled Castes in the Panchayat area, lived well away from the main village in which Thevars lived and which was where the Panchayat office was located. Besides this, a more significant factor was that the Paraiyars were the largest group of Scheduled Castes, and there had been conflicts in the past between Thevars and Paraiyars. The Thevars, therefore, preferred a woman from the Pallar caste. Arunthathiyars were a small Scheduled Caste group, but more importantly, the group was located in Endupalli village itself, and in their (Thevars) minds, the Gram Panchayat office was too easily accessible to them. Having someone living a fair distance away from the Panchayat office was preferable. It was easier to keep almost all Scheduled Caste people away from the Gram Panchayat office, which is what the Thevars wanted.

In Sivagangai district, the Thanjakkur Gram Panchayat has Thevars as the dominant caste. They control the functioning of the Panchayat. The only Scheduled Caste group in this area are the Arunthathiyars, and they comprise just four families. Thevars did not want any of the Scheduled Castes to contest the Panchayat elections or become the president, due to being Scheduled Castes, and also because there were only a few Scheduled Castes in the area, and it did not seem justifiable to have a president from such a small group. The Thevars had also threatened the Arunthathiyars and told them not to contest in the Panchayat elections or for the president's post. The Arunthathiyars were compelled to give a petition to the District Collector, that the post of president should be open to all, and not reserved. After an inquiry, the Collector ruled that the government order cannot be rescinded, and the post remains reserved for Scheduled Castes. While the Scheduled Castes were in a difficult situation due to the reservation of the president's post, the matter was solved in a manner of speaking, by one of the Thevars, a dominant individual, who said the Thevars should allow the Scheduled Castes to put up a candidate for the president's post. He ensured that "his" candidate would become president, though it was contested by another dominant Thevar, who had been the previous president. The former president had his son installed as the vice president, with the intention of running the Panchayat. This seemed to have been prevented by the former dominant Thevar who had "his" president in place.

The latter dominant Thevar indulged in malpractices, according to the president, and with the connivance of the Panchayat clerk. The president was illiterate, and could not directly stop these malpractices. He complained to the police, and the Thevar man was arrested. The district Panchayat suspended the clerk. The Thevar man then complained that the president had indulged in fraudulent practices, in the allocation of NREGS funds, and he (president) was then stopped from functioning as president. This was done at the time when the present study was being carried out. The police also did not accept the president's complaint that he and his family were being attacked by the dominant caste, i.e. the Thevars. Even when the president's young child was injured, they did not file any case against anyone among the Thevars who may have assaulted and wounded the child. The case against the president was in court, but even though the case had not been settled, the Thevars had threatened to have him dismissed as president.

The next Gram Panchayat that was studied is the Thiruvadanai Gram Panchayat in Ramanathapuram district. The largest number of residents of this Panchayat were Thevars and Yadavars. Presidents of this Gram Panchayat in the past were from one or the other of these two castes. Traditionally, it was the Thevars who were the leaders of this Panchayat, and there was considerable discontentment that the post of president was reserved for those of the Scheduled Castes. Accepting that they could not do anything about the reservation of this post, the Yadavars and Thevars decided to nominate a man from the Arunthathiyar caste, clearly with the intention of controlling him. A man from the Thevar caste was the vice president of this Gram Panchayat. Women of the Panchayat were elected with the support of their respective castes, but it was their husbands who functioned as the representatives, rather than the women themselves. The president initially had the support of different castes, including the Pallars, Thevars, Yadavars and Pilliamars. That enabled him to get elected. He was also a graduate, and had read about the Panchayats, and therefore, thought he could manage the Gram Panchayat without consulting the dominant castes of the area. In fact, for a while it looked as though he was not under the control of the dominant castes. However, that did not last, and they chose to oppose the development measures that he had started, raised false petitions against him, and filed a complaint against him with the District Collector. None of the petitions or complaints against him was found valid after due investigation, and all of them were dismissed. Notwithstanding this, the president is not able to function as the president because he is opposed by the Thevars. The Panchayat clerk does not work with the president, and finally, it should be said that his identity as a member of the Arunthathiyar caste has lost the support of even other Scheduled Castes such as the Pallars. The Yadavars who had been associated with the Thevars in bringing in this man as the president, now did not support him either, and thus, the president has been almost isolated in his functioning as a president, and is therefore unable to do anything in the Panchayat.

Members of the Scheduled Castes were not encouraged to participate in these Gram Panchayats, and this study has accessed information that indicates the participation and decision-making levels of the representatives of various groups in the Panchayats. The male representatives said they had some scope to participate in the Gram Panchayats, and the women representatives said they had less opportunity to participate in the Panchayats. In the case of the women, it was not only their lower caste which was the issue, but also gender considerations, which inhibit the women representatives from being able to participate. Women are dominated by various men, from their own families, their husbands, or dominant men of the villages. This hindrance was less visible with men, where it was their caste identity that came between their capacity to function more effectively in the Panchayats, but being men, they were not dominated by other men as in the case of the women representatives. However, when the representatives were asked whether the presidents of the Panchayats, who were all of the Scheduled Castes, could take decisions on their own, the almost unanimous response from both men and women representatives was an emphatic "no" (there was only one respondent who answered differently, out of the 54 representatives). Representatives who are Scheduled Caste members, and have been elected to the Panchayats through reserved seats, and quotas, can be seen to have been given an opportunity to be participants in the democratic institutions. However, as various studies, including this study, have indicated in other places as well, the level of participation in the Panchayats of members of the Scheduled Castes has been relatively low. This leads to the conclusion that for the most part, they are only nominal representatives, and not effective representatives of the Panchayats, through being denied the possibility of being participants in decision making in the Panchayats.

The role of caste identity in the Panchayats

Castes that are dominant in many areas of Tamil Nadu are often not "high" castes, but just above the Scheduled Castes in the traditional caste hierarchy. From what we saw in the areas dominated by different "dominant" castes, they were, first of all, able to play a decisive role in choosing who should be the president. That is to say, if the dominant caste did not make the choice, no one else got to be the president. We are talking about the Gram Panchayats that had the post of president reserved for those of Scheduled Castes. The clear understanding also was that the president would then function as a subordinate entity, under the direction of individuals of the dominant caste, or as it often happened, under the control of one dominant person who belonged to a dominant caste. The dominant individuals were able to wield more power not only because of their personal characteristics, but also because they were of the dominant castes. Both these factors together made them men of considerable authority and power, which was not only accepted by others, but also, rarely disputed. They were, effectively, leaders. Members of dominant castes do not accept such a condition in the Panchayats as the post of president being reserved for those of Scheduled Castes. Their belief is that someone "lower" to them in the sense of being a "low" caste cannot be placed above them in the Panchayat system, and as the president of a Gram Panchayat. The question of being lower to them appears to be a more sensitive issue at the village level, i.e. in the Gram Panchayats. From what we have seen of other Panchayats, at the higher levels, presidents of the Scheduled Castes do not usually face the strong antipathy and even the possibility of violence as those at the Gram Panchayat level. That is the reason that the dominant castes even contest the legitimacy of the system when the "low castes" are allowed to be in a position that puts them "above" those of the dominant castes, or other castes higher than the Scheduled Castes, even if they were not dominant castes. However, if the dominant castes are contesting the legitimacy of the Panchayats because of the reservation of the post of president for members of Scheduled Castes, we must also note that the lower democratic content in the functioning of the Panchayats, through the control of the dominant castes, and by reducing the participatory content of the functioning of representatives of the Scheduled Castes, also reduces the legitimacy of the Panchayats as institutions of democracy (on participation as a factor increasing or decreasing democratic legitimacy, see Benhabib, 1996). As we found, Scheduled Caste representatives, both men and women, were hindered and prevented from participating in the Panchayats and for the most part were not even allowed to speak in Panchayat meetings. A few representatives (men) said they could speak in the Panchayat meetings, but most of the representatives whom we interviewed said they did not speak, and the indication was that they were discouraged from speaking, by members of the higher castes, and dominant castes, and even if they were allowed to speak, their words were not taken seriously, or implemented (see also Dimitrov, 2018).

In all the cases of Scheduled Caste presidents, none was able to function independently, and was compelled to function under the direction of the dominant castes, and the dominant individuals of the Panchayat. When we considered even the presumably "non-caste" attributes such as higher education and land ownership, even though one of the presidents had a relatively higher level of education (graduate), and a moderate land holding, he was not allowed to function on his own as president. In fact, when he attempted to do so, he was immediately pulled down. This was after members of dominant castes had put up his name as president of the Gram Panchayat. Higher education and land holdings do not push up the status of a member of the Scheduled Caste to the point that the higher castes would accept his position as a Panchayat president, except as strictly nominal. In the case of women, Scheduled Caste members who had become presidents of the Panchayats, the fact of being a woman, and a Scheduled Caste, together pulled them down even more than a man who was a Scheduled Caste member and president. Their standing was such that the woman Scheduled Caste president was not even allowed to sit on the president's chair.

It is not just a matter of a higher caste claiming superiority in the social hierarchy. In Tamil Nadu, most dominant castes are also closer to the lower end of the caste hierarchy. They can be considered as "dominant low castes like the Thevars, Vellalars, Udaiyars, and Chettiyars. The members of Scheduled Castes considered these castes as higher castes" (Arun, 2007, p. 31). To the dominant castes, and to other castes as well, the legality of reservation is not the issue in having a "low caste" individual as president, but that their sensibilities have not been considered by the State, and that it has imposed on them what is not acceptable. Thus, what is often done is not outright opposition to a Scheduled Caste member being elected at all, but to ensure that he or she is not allowed to function as an effective president, and to convert the person to a purely nominal president. A different dominant caste was seen in each of these Gram Panchayats, but the ultimate result was the same. None of the dominant caste members said openly or explicitly that they were not in favour of having a Scheduled Caste president. They, instead, gave other reasons such as, that people of Scheduled Castes were employed by the dominant castes as wage labourers and agricultural labourers, and hence, it seemed most inappropriate that they were now, as presidents, occupying positions of some authority and importance. A Scheduled Caste president who functions effectively and independently as president is also not acceptable because the people of higher castes would then appear to be working under his "orders". Thus, even one who is better educated and of a higher economic level is not found as acceptable when he is of the Scheduled Castes, which is also defined as a "low" caste. It is also another matter that most times, people of Scheduled Castes also do not present a "common front" as it were, by joining together. We found that members of different Scheduled Castes did not join together to oppose the higher castes, and were only individual Scheduled Castes, and hence, smaller in number and easier to control.

Conclusion

Tamil Nadu is a state that shows a distinct antipathy to the practice of having presidents of Gram Panchayats who were members of Scheduled Castes. This does not mean that other states of the country do not to a greater or lesser extent show some degree of unhappiness to having those who the bulk of society (of each state) have considered as being at the "bottom" of the society, and therefore, not entitled to holding an important position such as being the president of the Gram Panchayat, or as sarpanch, or Gram Panchayat pradhan, whatever the post is called in different states. In most cases, the person (man or woman) who occupies the post of pradhan, or president as the case may be, is not given the respect due to someone holding that post, and the functions of the president are carried out by someone else, a man, of a higher caste, or a dominant individual who can control the Panchayat. The essential idea is that the person who is of a Scheduled Caste identity is not permitted to function as a president, and while in most places a Scheduled Caste member is able to get "elected", he or she is only a nominal president. The condition of Scheduled Caste women who hold these posts (president) is even more of a purely nominal status, because as women they also have the already existing reluctance of village society to permit women to function as a political representative. A man can function as a political representative, and in principle should be able to carry out these functions. When the job is done well, he is respected as an effective representative. If a woman becomes an effective representative, that is considered as being inappropriate, and in various states we have records of women being considered as of poor character, or loose women, because politics is not considered as an appropriate occupation or function of a woman (Vijayalakshmi & Chandrashekar, 2002; Tenhunen, 2009). The situation of the representative who is a Scheduled Caste member is of a status even lower from the point of view of being the incumbent of a political position. A woman who is of a relatively higher caste is not treated in the same manner as a Scheduled Caste woman is treated, even one who is a Gram Panchayat president, who is not given any respect at all as we saw in Tamil Nadu.

The dominant lower castes controlled the functioning of the Gram Panchayats when the president's post was reserved for a member of a Scheduled Caste. They were able to do this due to their reasonably large numbers in the village, their land ownership, and also the more aggressive projection of their political power in the village. Members of Scheduled Castes, even if they were of a larger number in some villages, were not able to project such power (even with their numbers), were more dependent on the locally dominant group through the need to find employment with them, as agricultural labourers, and also the fact that most often, they were not a compact, well-knit group, in the sense of being of one Scheduled Caste group. They were individual castes, and at various times were even seen as not being supportive of a president who was of a Scheduled Caste, but a different caste. Hence, they were easier to overcome in terms of the dominant castes getting their own way, in selecting a person who would be the president, or in controlling the position of the president, either through directly functioning as the *de facto* president, or instructing the president to do what they tell him/her to do. The reservations have no doubt brought people of the Scheduled Castes into the Panchayats, as representatives and as presidents, but have not been seen to provide them with effective power to function either as representatives, or as presidents. Since the Gram Panchayats are at the local level of villages, dominant groups and individuals are able to control the functioning of the Panchayats, and the Scheduled Caste individuals, who still have the status of being the lowest in the caste hierarchy, through generations of being downtrodden, leaving them in a lower economic condition; all these points have not been overcome with the enactment of the 73rd Constitutional Amendment. Even though it gave certain groups a modest opportunity for social mobility for the most part that has not materialised, at least through the Gram Panchayat.

Finally, what we can state is that the position of the president of the Gram Panchayat confers power and prestige on those who become presidents, if they are from locally dominant castes, or castes higher in the caste hierarchy.

However, if the person is from a Scheduled Caste, the position of president does not confer any power or prestige, but what people see is the caste identity of the incumbent president, rather than the position of president. Thus, the Scheduled Caste identity takes overwhelming precedence over any other marker of identity that the person may possess, for example wealth or higher education, or even both together.

References

Arun, C. J. (2007). *Constructing Dalit identity*. New Delhi: Rawat Publications.

Benhabib, S. (1996). Toward a deliberative model of democratic legitimacy. In S. Benhabib (Ed.), *Democracy and difference: Contesting the boundaries of the political*. Princeton, NJ: Princeton University Press.

Chandrasekhar, L. (2011). *Undermining local democracy: Parallel governance in contemporary South India*. New Delhi: Routledge.

Dimitrov, B. E. G. (2018, August, October). A case study on the performance and problems faced by women village Panchayat presidents in Vadamadurai Block, Dindigul District, Tamil Nadu. *Pune Research Discovery*, *3*(3).

Frontline Magazine. (2013, April 19). Print edition: Disempowered dalits.

Inbanathan, A., & Gopalappa, D. V. (2003). Fixers, patronage, 'fixing', and local governance in Karnataka. *Sociological Bulletin*, *52*(2), 164–185.

Kannan, T. (2007). *Formation of Dalit identity among the 'untouchables' of Tamil Nadu: A sociological study from a human rights perspective* (Unpublished PhD dissertation submitted to Osmania University, Hyderabad).

Lakshman, N. (2011). *Patrons of the poor: Caste politics and policy making in India*. New Delhi: Oxford University Press.

Leiten, G. K. (1996). Panchayats in western Uttar Pradesh: 'Namesake' members. *Economic and Political Weekly*, *31*(39), 2700–2705.

Leiten, G. K., & Srivastava, R. (1999). *Unequal partners: Power relations, devolution and development in Uttar Pradesh*. New Delhi: Sage Publications.

Karthikeyan, D., Rajangam, S., & Gorringe, H. (2012). Dalit political imagination and replication in contemporary Tamil Nadu. *Economic and Political Weekly*, *47*(36), 30–34.

Martin, N. (2015). Rural elites and the limits of scheduled caste assertiveness in rural Malwa, Punjab. *Economic and Political Weekly*, *50*(52), 37–44.

Mathew, G. (2003). Panchayati raj institutions and human rights in India. *Economic and Political Weekly*, *38*(2), 155–162.

Mathew, G., & Baviskar, B. S. (Eds.). (2009). *Inclusion and exclusion in local governance: Field studies from rural India*. New Delhi: Sage.

Mayer, A. C. (1958). The dominant caste in a region of Central India. *Southwestern Journal of Anthropology*, *14*(4), 407–427.

Radhakrishnan, P. (2002). The Vanniyar separatism. *Frontline*, *19*(17).

Srinivas, M. N. (1959). The dominant caste in Rampura. *The American Anthropologist*, *61*(1), 1–16.

Tenhunen, S. (2009). *Means of awakening: Gender, politics and practice in rural India*. Kolkata: Stree.

Vijayalakshmi, V., & Chandrashekar, B. K. (2002). *Authority, powerlessness and dependence: Women and political participation* (Working Paper 106). Bangalore: Institute for Social and Economic Change.

Vincentnathan, S. G. (1996). Caste politics, violence, and the Panchayat in a South Indian community. *Studies in Society and History*, *38*(3), 484–509.

4

CASTE AND THE BUSINESS OF DEMOCRACY

Insights from rural Karnataka[1]

Kripa Ananthpur and Radhika Viswanathan

Introduction

Over the past ten years, Panchayat budgets in Karnataka have seen a staggering increase – from fivefold to fifteenfold. This is a significant departure from the norm as one of the recurring problems of decentralisation in India has been the poor devolution of funds to the lowest tier of governance – Gram Panchayats. Decentralisation literature is replete with studies that show how functional devolution has not been commensurate with fiscal devolution, leading to inefficiencies in service delivery at Gram Panchayat level (Rao et al., 2011). Studies on the impact of increased devolution of grants to Panchayats are limited to the functional efficacy framework and do not extend to understanding its impact on local democracy. Hardly any research has been done on how Gram Panchayats behave when the state *does* devolve adequate funds to the lower tiers of governance. The impact of increased devolution of funds to local Panchayats on rural society has not met with deeper inquiry.

The role played by caste in politics in India has been extensively studied (Srinivas, 1955; Kothari, 1970; Chhibber, 1999; Jaffrelot, 2003; Chandra, 2007) but there is very little exploration of the impact of increased fiscal devolution on political contestation and caste at the Panchayat level. The term "votebank" itself was coined by M. N. Srinivas in his well-known essay (Srinivas, 1955), linking a community's primal sense of obligation and a powerful patronage system between the electorate and the candidate. Well before the 73rd Amendment, these primordial connections played themselves out during state and central elections (Srinivas & Shah, 2007) and subsequently by highlighting the influence and interconnectedness of state and central level politics on local politics and effect of "ethnic voting" on national elections (Jaffrelot & Verniers, 2009). Closer to this study, the presence of votebanks in Karnataka points not only to the presence but evolution of "obligation and reciprocity" over time (Breeding, 2007). Within the decentralisation discourse in particular, caste literature has largely focused on ethnicity and political mobilisation, and how they support Indian democracy (Chandra, 2007), political reservations and favourable outcomes for minorities (Duflo, 2005).

The impact of decentralisation after the 73rd Amendment has been studied extensively (Chattopadhyay & Duflo, 2004a; Manor, 2007; Mansuri & Rao, 2012), as has the allocative efficiency of political institutions and caste and sub-caste power structures. Public resources are often used to support private interests, benefiting the groups represented by elected representatives but also by voters, who access these resources using influence (Besley et al., 2007). However, while the decentralisation process does create spaces for marginalised communities to participate in public resource allocation (Besley et al., 2011), albeit imperfectly, Palaniswamy and Krishnan's work on Panchayats in Karnataka found that village councils are prone to "elite capture" and in some cases, by dominant castes resulting in the inequitable allocation of public resources (Palaniswamy & Krishnan, 2008, 2012).

There has also been a considerable focus in the decentralisation discourse on issues related to fiscal devolution, on caste, gender and public good allocations. Reservations can help groups such as women access resources and participate better (Chattopadhyay & Duflo, 2004b). However, broad caste categorisations do not often capture intra-caste or *jati* dynamics, which often reflect a more accurate lived experience and can have significant implications when it comes to resource allocation (Joshi, Kochhar, & Rao, 2017). The continued presence and the synergy between customary institutions and Panchayats indicate that caste and caste-based institutions are adapting and working with local governance structures (Ananthpur, 2007; Ananthpur & Moore, 2010). The impact of affirmative action and decentralisation has been extensively studied by scholars. While the effectiveness of reservation of participation and governance outcomes has been contested, there is general concurrence on the importance of mandated spaces for political participation (Nayak, 1995; Osborne, 2001; Omvedt, 2006; Inbanathan, 2003; Inbanathan & Sivanna, 2010).

While the role of votebanks in influencing electoral outcomes at the state and national level elections has been extensively studied, there has been little research on the influence/role of ethnicity or votebanks at the local Gram Panchayat (GP) elections post–73rd amendment to the Constitution (Chattopadhyay & Duflo, 2004a; Dunning & Nilekani, 2013). Further, there is even lesser attention paid to the complex interplay of increased fiscal devolution, local elections and votebanks during GP elections. For instance, how the increased fund flow is challenging assumptions on votebanks and caste configurations or the way different caste groups respond to increased devolution to local governments has not been systematically documented.

In attempting to address these questions, this chapter makes an original contribution to the decentralisation discourse by examining fiscal devolution alongside voting behaviour and caste at the local level.

Context

Villages in Raichur, Gulbarga (and parts of now Yadgir district) are amongst the more backward regions of Karnataka where consistently poor human development indicators, per capita income and an overall lack of growth have made these some of the most poorly performing districts in the state. Still bearing the burden of historical institutional and administrative legacies, and unlike many of the southern ex-Mysorean districts that enjoy the benefits of a legacy of progressive development policies and the *ryotwari* system (Govt of Karnataka, 2002), in Gulbarga and Raichur the colonial legacy, and vestiges of old social patronage networks (such as the *jagirdari* system) still determine the socio-political constructs of daily life and have slowed the adoption of a range of investments in critical areas like agriculture, education and health, negatively impacting literacy and health levels (Banerjee & Iyer, 2005).

Even today, Gulbarga and Raichur fare poorly on almost all indicators. Out of 30 districts in Karnataka, for both HDI and GDI, they are ranked at the bottom of the list: Gulbarga is 26th, and Raichur is 27th in the 2005 state HDI rankings (Govt of Karnataka, 2005). Literacy rates are well below the state average (75.60 per cent) at 65.65 per cent and 60.46 per cent for Gulbarga and Raichur respectively. Gender indicators are poor as well, with female literacy in Gulbarga at 55 per cent and 48.73 per cent in Raichur and poor sex ratios (949 for every 1000 males in Raichur and 935 in Gulbarga). There are also several instances of maternal and infant mortality due to limited physical access to health facilities in these regions (Krishna & Ananthpur, 2013). Overall, per capita income is low as well and although Gulbarga, Raichur and Yadgir are amongst the specially designated backward districts earmarked for priority interventions (under the 118th Amendment Bill, 2012), they remain industrially backward districts struggling with poor growth.[2]

The villages included in our research in Raichur and Gulbarga (and now Yadgir), are representative of this "backwardness". Our research was conducted within a larger intervention that was being tried in Karnataka between 2007 and 2012, to see if induced civic participation in a backward region in Karnataka could lead to deepening of democracy and in turn better governance outcomes. The larger intervention to deepen democracy through inducing people's participation in rural Karnataka, India was conducted by the World Bank, within the World Bank aided Gram Swaraj project in Karnataka. The research was designed as a mixed methods study, using both quantitative and qualitative tools over a period of five years (2007–2011).

A qualitative sample of a subset of 26 villages from 10 Gram Panchayats was identified for detailed, intensive ethnography. These villages had a mixed caste and community profile: largely Lingayats, Kabbaligas, Kurubas, Kumbars, Scheduled Castes (SC), Scheduled Tribes (ST) and Muslims. A few Brahmin families were present in one Panchayat.

In all the villages, agriculture was the primary occupation, with paddy, sunflower, jowar and gram being the primary crops. Other agriculture-allied occupations were wage labour and livestock. These villages experienced some seasonal migration to larger cities in search of work and very few instances of reverse migration. Schemes like the National Rural Employment Guarantee (NREGA) have brought in livelihood support, and other housing schemes (Indira Awas Yojna, the Ashraya Housing Scheme) and the Backward Regions Grant Fund were also active. In terms of HDI, the villages figure in the lower end of the Panchayat and village HDI rankings for Karnataka (Government of Karnataka, 2015).

While overall these villages may not manifest a highly skewed distribution of land, feudalism, rent seeking, patronage and nepotism persist through the erstwhile *jagirdars*, who are now influential families in the villages (Ananthpur, Malik, & Rao, 2014). However, the extent of their influence varied over the period of our study. The research area was also fraught with undercurrents of violence. In many villages, there have been numerous instances of theft, threats, dacoity, smuggling in some of the more "frontier" villages near the Maharashtrian border, and even murder.

Over the past 15 years, these villages have also seen a sharp rise in the quantum of fiscal devolution. The increase in budgets, coupled with limited economic opportunities apart from traditional occupations and low levels of human development, has created a simmering but latent undercurrent of discontent in the villages. This latent "discontent" permeating these villages tends to get exacerbated by different triggers, and local elections is one of them. These flashpoints bring this latency to a head, transforming the way people participate in local elections, identify caste affiliations and votebanks and use it to their individual and community benefit.

Methodology

A mixed method research was carried out to study the impact of induced civic participation on local democracy in Raichur and Gulbarga districts of Karnataka. The qualitative part of this research was carried out in 26 villages belonging to 10 Gram Panchayats in Gulbarga and Raichur districts (see Table 4.1). The qualitative study relied on an array of research methods and tools that included ethnography, direct observation, in-depth interviews, focus group discussions and so on. The qualitative research thus elicited extensive, detailed documentation of village life in the villages. Between 2007 and 2011, we carried out intensive ethnography where our

Table 4.1 List of Gram Panchayats and their caste composition

Panchayat	*District*	*Dominant castes*
Maranahalli	Raichur	Lingayat, SC, Muslim
Utsavahalli	Raichur	ST and SC
Aalanahalli	Raichur	Lingayat, SC, ST, Nayak
Artihalli	Raichur	Nayak, Lingayat
Kirkhalli	Gulbarga	Lingayat, Muslim, Kuruba, SC
Dangehalli	Gulbarga	Lingayat, Reddy, Muslim
Dharmapuri	Gulbarga	SC, Lingayat, Muslim
Iranahalli	Gulbarga	SC, ST, Lingayat
Jhakuri	Gulbarga	Lingayat, SC
Kotehalli	Gulbarga	Kabbaliga, Kuruba, Lingayat, SC

Source: Author.

field investigators either stayed in the village or visited the village every day and documented the everyday life of the citizens. Panchayat meetings, Gram Sabhas, village development works, conversations at the village centre, celebrations such as festivals and marriages, everyday lives of Panchayat members, ruminations of village elders, village factionalism and so on were observed and documented.

A series of elections that were held during this period such as state (2008), national (2009) and local (2010) were documented and extensive interviews with all the contesting candidates were recorded. We had the advantage of having collected election data for two terms, 2010 and 2015, through direct observation. Factual data on those that won and lost the elections was revalidated through the Panchayat office. But since we entered these villages after the 2005 Panchayat elections had been conducted, our 2005 election data is both based on a recall as well as factual data collected from the Panchayat office.

The investigators were asked to track electoral activities and expenditures, examine changes in levels of local activism, and investigate other issues that were relevant to the political and economic life of the Gram Panchayat (Ananthpur et al., 2014). From 2012 to 2017, our investigators made periodic visits to the villages and updated their data. Pre- and post-2015 election phases were covered with interviews of candidates contesting for GP elections.

The qualitative research thus elicited extensive, detailed documentation of village life from 26 villages in the northern Karnataka districts of Raichur and Gulbarga. A preliminary analysis of some of the data on local elections began to display interesting patterns in terms of caste alliances and caste dynamics. This led us to examine the phenomenon of caste dynamics in Panchayat elections in the context of increased devolution of funds to rural local bodies in greater depth. What we report here is a set of preliminary research findings limited to the districts of Gulbarga and Raichur and not the rest of Karnataka.

Caste dynamics in Panchayat elections

Rajni Kothari's (Kothari, 1970) seminal work on the analysis of the relationship between caste and politics in India provides a comprehensive view of the ways in which caste groups behave in response to electoral competition. Further, M. N. Srinivas's analysis of this caste dynamics through the "votebank" lens illustrates how the patrons or village *influentials* engage in reciprocal transactions with political leaders by delivering votes through the votebanks in exchange for economic and other kinds of favours (Srinivas, 2002). Although these analyses were carried out before the 73rd amendment to the Constitution that changed the nature of polity at the local level by bringing in affirmative action, this framework of analysis is still relevant today. Our field findings not only confirm but take it further in the context of local government elections.

The preliminary analysis of the pre-election data threw up two broad, interesting findings that we observed across our 26 villages.

i. Increased contestation

The first general finding that we report here relates *to a steady increase in contestation* in GP elections. This is a trend observed across Karnataka. For example, in 2005, there were 74,496 seats up for election, and 1,38,215 candidates in the fray.[3] In 2010, the number of candidates increased, and approximately 2,32,000 candidates stood for election for 80,159 seats[4] and in 2015, there were 2,33,493 candidates for 94,458 seats.[5] Across our 26 villages, we noticed that the number of contestants to GP elections had risen significantly as well: In 2005, 241 candidates stood for election for 120 seats; this rose sharply to 355 candidates for 128 seats in 2010, and 348 candidates for 139 seats in 2015.

In general, the number of candidates per seat in the Gram Panchayat elections has greatly increased. Where elections used to be fought between a few candidates or candidates were elected unanimously, the ratio of candidates to seats has increased exponentially. For example, in Kirkhalli, during the 2010 Gram Panchayat elections, 42 candidates stood for 18 seats. In Aalanahalli, 50 candidates campaigned for over 12 seats. A similar number was seen in Utsavahalli where 52 candidates campaigned for 14 seats. Dangehalli saw 35 candidates fighting for 11 seats; 23 for 9 seats in Maranahalli, 28 for 8 seats in Iranahalli, 33 for 14 in Artihalli, 38 for 18 seats in Jhakuri and 35 for 11 seats in Dharmapuri.

We also noted instances across all Panchayats of increased youth interest and aspiration to stand for election, and that village level forums such as

School Development Monitoring Committees (SDMC) are being used by aspirants as a stepping stone to contest Panchayat elections. Village youth are forcing the village elders and other political leaders to allow them to contest GP elections. There have been several instances of youth participation in Gram Panchayat elections.

If we further delve into this to see which social groups contested elections (see Table 4.2), we find a significant increase of Scheduled Caste (SC) candidates contesting in Panchayat elections, expressed both in absolute numbers and as a percentage of total candidates contesting for elections. For example, between 2005 and 2015, the percentage of SC candidates increased from 23.2 to 31 per cent (in absolute terms, this is 56 SC candidates out of 241 in 2005 to 108 SC candidates out of 348 in 2015). In the same period, we find a decrease in Other Backward Castes (OBCs) contesting elections, from representing 56 per cent of all candidates in 2005 to 48.5 per cent in 2015. Other social categories, Scheduled Tribes (ST), and other communities' participation has remained more or less constant in the same period. This affirms the point that more SC candidates seem to be contesting Panchayat elections.

ii. Increased contestation for general seats

The second general finding across the sample villages was a significant increase in contestation for general category seats. An even more interesting finding was that a significant percentage of this increased contestation for general seats came from reserved categories such as SCs.

A further disaggregation of the data shows that apart from contesting for reserved seats, an increasing number of SC candidates are contesting for general seats (see Table 4.3). This trend is seen both in terms of absolute numbers and as a percentage of total SC candidates. In 2005, 9 per cent of SC candidates stood for general seats, and this increased to 11 per cent in 2010. In 2015, 20 per cent of all SC candidates contesting GP elections stood for general seats. Interestingly, not only SC candidates are contesting

Table 4.2 Percentage of SC, ST, OBC and other communities contesting GP elections (2005, 2010 and 2015)

	2005	*2010*	*2015*
Total SC	56 (23.2%)	98 (27.6%)	108 (31%)
Total ST	44 (18.2%)	62 (17.5%)	63 (18.1%)
Total OBC	135 (56%)	187 (52.7%)	169 (48.5%)
Total other communities	6 (2.5)	8 (2.2%)	8 (2.3%)
Total seats	241	355	348

Source: Election rosters for 2005, 2010 and 2015, collected from the Gram Panchayat offices.

Table 4.3 Breakup of SC candidates across seats

GP Election cycle	*% of SC candidates standing for general seats*
2005	9%
2010	11%
2015	20%

Source: Election rosters for 2005, 2010 and 2015, collected from the Gram Panchayat offices.

in the general category but a small number is also winning. While in 2005 one SC woman contesting under the general category seat was unanimously elected in one GP, this figure went up to three in 2010 (through contested elections) and by 2015 five SC candidates had won General Category seats through contested elections from four GPs. The increased contestation in Panchayat elections that we observed manifested itself in several ways at the village level.

a) Intra-caste competition

The increased intra-caste competition we witnessed in northern Karnataka districts was not only within "entrenched and ascendant" castes (Kothari, 1970) but also within SC caste groups.

The intense contestation within caste groups was visible not only across villages but also across all caste categories in our sample villages. This fissiparous nature is further manifesting itself in intra-caste and sub-caste (jati) competition. This was witnessed in six out of our 10 sample GPs. We illustrate this with some examples below.

Pinjar vote split in Maranahalli

Maranahalli Gram Panchayat is a small Panchayat in Raichur district. The main village, Maranahalli, has around 840 houses with roughly 4,000 people.[6] It is a multi-caste village, mainly populated by Lingayat, SC and Muslim families. Muslims constitute two-thirds of the village population. The primary occupation in Maranahalli GP is agriculture, with paddy, sunflower and jowar being the dominant crops. A significant proportion of the population, mainly SCs and Muslims, are also engaged in agricultural labour. In Maranahalli village alone, out of the main working population, 49 per cent are main cultivators and 31 per cent are agricultural labourers.

Civil society is vibrant in Maranahalli with several SHGs and youth groups being quite active in the village. The traditional/

customary Panchayat is active in organising religious life within the village and also maintains internal law and order with the help of community elders like Ramagowda and Shankargowda, or Ramalingappa and Husseinsaab who help settle disputes in the village.

The village is predominantly divided between two political parties, with people choosing between the Janata Dal (Secular) or the Congress party, but these affiliations are only activated during state or central elections. Participation in local governance matters is restricted to Panchayat members and there is little active participation by villagers.

Maranahalli has three GP seats allocated to it, out of which one seat was reserved under BCA category in 2005 GP elections. Village elders asked the Muslim community to discuss among themselves and choose the appropriate candidates to contest. Extensive debate took place among the Muslim leaders and the community regarding the nomination of candidates. The contest was finally narrowed down between Husseinsaab and Kaleemsab Pinjar. Husseinsaab had a good reputation in the village not only among the Muslim community but also among other caste groups. Hence the elders decided to elect him unopposed. Hearing this, Kaleemsaab decided to abide by the decision of the leaders as he felt that going against them would cast him in a bad light in the village and also his chances of winning would be slim without the support of the elders. Thus in 2005, Husseinsaab was elected unopposed to the BCA seat in Maranahalli.

However, by the time the 2010 GP elections were announced, the mood in the village had changed. The election was being held within the context of higher devolution of funds to the Panchayats. There was intense competition to contest for elections. Unlike in the previous term, the competition within the Muslim community was fierce this time. Maranahalli has two sub-groups within the Muslim community – Pinjar which has around 350 households and Sheikh sub-community which has only 10 households. The people belonging to the Sheikh community felt that they had been sidelined in the previous elections as the seat had been unanimously given to Husseinsaab who was a Pinjar. They felt that this time, a representative from the Sheikh community should be given a chance in the GP. Some representatives belonging to the Sheikh community approached the village elders and asked for a chance for representation to be given to the Sheikh community. The elders felt that this was a fair request and were favourably inclined to support the candidature of Ibrahimsaab. However, this was not acceptable to the Pinjar community as they felt that they were the majority population and had the right to contest. Further, unlike in previous terms, there was no effort by the Pinjar community leaders to come out with a consensus candidate as several individuals were keen to contest. Gradually, there were eight candidates for the BCA seat in Maranahalli. From the Sheikh community, not only Ibrahimsaab filed his papers but also his elder brother and supporter Abdulsaab filed his nomination as a backup in case Ibrahimsaab's papers were rejected. Meanwhile, from the Pinjar community, people started filing nominations indiscriminately. When candidates from three influential Pinjar families filed their nominations, a fourth man, T. Pinjar, told our investigator, "Why should I not file a nomination. They have five families each and I have more families' support". So, he also decided to stand for elections. These developments prompted the fifth candidate, B. Pinjar, to enter the fray as he felt it was a prestige issue as otherwise he would be considered inferior to his relatives for not contesting the elections.

A sixth Pinjar (a sibling of one of the other candidates), K. Pinjar, also decided to stand after his sponsor Suryappa convinced his father by saying, "look, your boy is an educated person. Why should you not make him stand for election? This time, it is sure that the president's post will get reserved for BCA. If your boy files his nomination, we will all give him full support. Your community will be benefited and it will also be good for the whole village." Thus K. Pinjar also filed his papers.

In this manner, for one seat (reserved) of BCA, there were eight contestants – all from the Muslim community. There was no coordination or consultation among them. They had also not sought the support of the elders. Barring a couple of candidates, the rest of the candidates did not have any support base in the village. Further, these candidates belong to the same Muslim family circles – cousins, uncles, nephews, etc.

Such indiscriminate filing of papers by Muslim aspirants irked the Muslim community elders and leaders. They thought that intra-community fights like this were undesirable, and they set in motion consultations and negotiations by calling for meetings of all the concerned parties and the voters. They continued efforts till the last day of withdrawal of nominations. Despite hectic negotiations, the elders were unable to arrive at a consensus candidate. They were successful in getting three candidates to withdraw their nomination papers – two withdrew in favour of the Sheikh candidate and one in favour of a Pinjar candidate. This left five candidates in the fray for one seat. Muslim community leaders were upset at the lack of unity among the community members as they felt that such discord within the community was shameful, especially as Lingayat community members had arrived at a consensus candidate from their community for another GP seat without displaying such public dissonance.

The Pinjar infighting was seen as even more shameful as they were all related to each other and belonged to one large, joint family. The Pinjar infighting was further exacerbated with the two political rivals in the village, one belonging to the Lingayat community and another to a SC group, both backing two different Pinjar candidates.

Finally, during the GP elections, due to the infighting and division of Pinjar votes, Ibrahimsaab won the elections under the BCA seat. Ibrahimsaab who runs a PDS shop is well liked by the villagers as he is known to give rations such as kerosene, rice and sugar to people even if they are unable to pay immediately. He is also known to financially help those in need. However, some people also alleged that Ibrahimsaab had used his goodwill and money to incite his Pinjar friends to contest elections, thus dividing the Pinjar votes and winning the election.

The Muslim community leaders felt that this public display of discord had shamed the community in the eyes of fellow villagers. Hence when the 2015 GP elections were called, they were determined to present a united front.

Since the seat was reserved for general women and not BCA in 2015 elections, the Muslim community leaders requested the village elders that "despite being numerically dominant, we were unable to maintain unity and come out with a consensus candidate. This time we will not repeat our mistakes. Please give us a chance for representation." The Pinjar community unanimously proposed the name of Fatima Begum, wife of Numaansaab, as the candidate under the women's general category. Since the village elders and others supported this move, the Sheikh families decided to support Fatima Begum and did not propose an alternative candidate. Hence the 2015 GP elections were conducted peacefully.

A similar story of intra-family election feud was documented from an SC family in Dangehalli, where five members of the same family expressed interest in standing for elections. Equally fierce has been the competition among sub-caste (jati) groups to contest under the reserved category for 2010 elections in Utsavahalli. For example, in one case, we documented intra-caste fighting for the SC-women seat – between Mala, Madiga, Waddar and Dasaru sub castes.[7]

Such stories of intra-caste and sub-caste competition were not outliers and were also documented in some of our other sample villages.

b) Inter-caste cohesion

In response to this increasing fragmentation of caste as a votebank at the local level, caste leaders constantly attempt to keep the caste cohesion intact. The Pinjar story presented in the previous section aptly illustrates this point. These attempts, being carried out against the political canvas within the village are constantly thwarted (or attempted to thwart) by rebels within the same caste group or even from those belonging to other caste groups that would benefit from the fragmentation of these votebanks.

For instance, the reference made by the Pinjar leaders to the ability of the Lingayat community to arrive at a consensus candidate was not a simple, straightforward exercise as the Muslim elders believed it to be. Rather it was fraught with intrigue and attempts to break the caste cohesion by vested interest groups.

Lingayat unity under threat: SC leader's failed move to thwart Lingayat community's attempt at consensus

In Maranahalli, the various Lingayat families were trying to avoid dissent and were trying to unanimously elect a Lingayat candidate. At a community meeting, Ramagowda said,

If you give support to some other party, there will be no coordination and understanding in our Lingayat community. It is a matter of the honour of our community – not an individual's case. Therefore all should unite and act. In other times, you can have your own views and say – we will not come in your way – that freedom is always there for you. In our community, there are many political leaders and politicians. Many officers of government are known to us. So we should ensure that we win the GP seat.

Yet, others were trying hard to break this unity. Mallikarjuna Chalawadi (SC) and Ibrahimsab approached Ramalingappa Julad (Lingayat) and told him, "What is the use of always sermonising that all are belonging to one-caste, one family, brethren etc. Is it fair that every time they grab the power? Do you not deserve it? Why should they deprive you of this right? You are in no way inferior or less to those fellows; you have enough support from many families. Why should you be afraid and why should you withdraw nomination of your wife?"

However, these efforts were unsuccessful and thanks to hectic negotiations among the Lingayats, the unanimous election was successful.[8]

Similarly, in Utsavahalli, multiple candidates – five candidates – were trying for one SC seat and attempts were made by village leaders Gundappa Gowda (Lingayat) and Syedsab (the Taluk Panchayat member) to avoid a split in SC vote (between Waddar, Madiga and Chalawadi). This was important not only to the SC caste groups but also other caste groups within the village as they did not want an internal vote split between various sub-castes to impact the electoral outcome in favour of one of the candidates – Ganesh Murthy – the "corrupt" husband of the former GP president.

Ustavahalli: engineering unity at any cost

In Utsavahalli Gram Panchayat, hectic negotiations were on to ensure that Ganesh Murthy, the ex-GP president's husband would not win the SC seat in 2010 GP elections. Ganesh Murthy, who was the *de facto* power behind the 2005 GP president (his wife), had earned a bad reputation within the village for his excessive corruption and extraction. Villagers were unanimous in their desire to ensure that he did not return to power.

Five candidates were contesting for the reserved seat: Ganesh Murthy (Madiga), Ramaiah (Waddar), Basavaraj (Waddar, the Huligamma temple priest,), Nagamma (Madiga) and Mallappa (Chalawadi). Ganesh Murthy's opponents – Taluk member Syedsab, Lingayat leader Gundappa Gowda, V. Nayak and other BJP leaders and other opponents – met and decided Ramaiah was the strongest candidate to contest against Ganesh Murthy. The others would have to withdraw to avoid splitting the SC vote in Ganesh Murthy's favour.

It was understood that the other contestants would be compensated for the electoral expenses incurred. Except for one candidate who allegedly wanted more compensation, all others withdrew their nomination papers in favour of Ramaiah.

Our investigator documented that early on the last day for withdrawing nominations, Ganesh Murthy went to Ramaiah's house at 5.30 a.m. and offered him money to withdraw his nomination. Ramaiah refused, saying, "Come what may, even if you weigh me in currency notes or bury me in garlands of notes, I will not agree." When news of this reached the rest of the village, 20–30 young supporters of Ramaiah gathered outside his house and demanded Ramaiah not give in. Ganesh Murthy, desperate to win, offered Ramaiah an even higher amount but Ramaiah did not relent. In the end, Ramaiah won the seat in April 2010.

c) From voting to contesting: the shifting nature of traditional votebanks

Kothari (1970) refers to the way lower caste groups and those that were traditionally not part of the political process were used to garner support by the "ascendant or dominant" caste groups in the interplay of caste and politics. However, Srinivas had foreseen the potential impact of secret balloting on these votebanks. Even in 1952, he found that the candidate that had the support of the village headman had been defeated in the election. It was alleged that the Harijan clients had "double-crossed" him and voted for the rival (Srinivas, 2002, p. 448). What we are witnessing today in north Karnataka villages is the further weakening of these ties. This is evident in the way in which these "support" groups are rebelling against being mere votebanks and aspiring to contest elections on their own merits. For example, in Aalanahalli, Kurubas contested for the first time in 2010 to get benefits for their own community.

> According to the Kuruba community in Aalanahalli Panchayat, they are interested in contesting elections because until now all the 'benefits' – government projects and funds – have gone to other communities. They feel that people from other communities such as Shivanna Nayak, who used to be poor and whose family used to work as coolies in the fields, are now financially sound. Others like Gundamma have been able to construct houses.[9]

Similarly in Kotehalli, caste groups that were traditional vote banks for other caste groups have now started asserting their rights to contest local elections.

> *A bitter taste in the mouth: votebanks assert themselves against traditional patrons*
>
> Kotehalli has sizable Kabbaliga and Kuruba populations and a smaller Badiger community. Neither the Kabbaligas (who are the majority community in Budur village) nor the Badiger community has been represented at the Panchayat for the past ten years. They have always supported the local Dalapati's[10] (a traditional position designated to maintain law and order in the village) choice for candidate.
>
> When the Dalapati (a Kabbaliga) died, there was a lot of interest from the Gowdas to take up the Dalapati's post. Basavarajgouda, a local leader with political aspirations, schemed to make Seshappa, the son of the deceased, the next Dalapati. He did this in the hope of earning Kabbaliga support at the next GP elections in 2010.
>
> Ironically, at election time, the Kabbaligas realised that other communities were benefiting from being part of the local government, and in a surprise twist, they decided to field their own candidate, Sharanappa, against Basavarajgouda. The Kabbaligas offered Basavarajgouda their support if he decided to field his wife for the women's seat. Basaravajgouda refused. They then struck a

> deal with the Badiger community to support each other's candidate if the Badigers would field Devi, Srinivas Badiger's wife, for the women's seat.
>
> As one villager explained to our investigator, "The Kabbaligas now suddenly realised that due to dividing their loyalties, they were not getting any benefits for their community". Therefore, they united and even Seshappa had to be part of the community decision.
>
> Basavarajgouda tried to break the Kabbaliga–Badiger unity by winning over Badiger support. He thought that by getting Devi's opponent (fielded by the Lingayats) to withdraw her nomination, ensuring Devi's unanimous election, they would vote for him (for the general seat). When Devi's opponent withdrew, "the Badiger leaders such as Srinivas Badiger and others allegedly celebrated the event by drinking and dancing all night". However, Basavarajgouda did not anticipate that there would be other candidates standing for the women's post, and in the end, the elections for the women's seat went ahead as there was a second candidate. In the end, Devi won the women's seat and Sharanappa won the general seat. Basavarajgouda did not win and this defeat was, as villagers say, "like a bitter taste in the mouth".[11]

However, these emerging caste dynamics that we observed were largely limited to local elections and were not witnessed to the same extent at the state and national elections. The behaviour of these same caste groups during state and national elections were more predictable where interests converged either on political lines or caste lines.

Although the reconfiguration of caste around electoral politics witnessed in northern Karnataka villages is reminiscent of Kothari's (1970) analysis of the relationship between caste and politics, this research brings in the added dimension of the role of increased devolution of funds to this framework of analysis to explain this new trend in local elections.

Money matters

The period between 2005 and 2010 witnessed a significant rise in the devolution of funds to rural local bodies in Karnataka, both through state and central transfers. The National Rural Employment Guarantee Scheme (NREGS/NREGA) as a tied grant contributed to a significant increase in funds being devolved to rural local bodies (RLBs).

Given below is an example of how various schemes with both tied and untied funds have led to a huge increase in funds being devolved to the Gram Panchayats. The annual grant for each GP in Karnataka is around Rs 6 lakh (Rs. 0.6 million). In addition, they receive grants tied to various programmes. NREGA budgets are based on a labour component and on average GPs in Karnataka received nearly Rs 30–35 lakh as NREGA grants. In addition, during the period 2005–2010, Backward Region Grant Fund and Suvarna Gram Yojna programme contributed additional resources to the hitherto cash-starved Panchayats of northern Karnataka which have been traditionally poor in raising local resources. BRGF grants for backward regions constitute around another Rs 6–7 lakh per annum. However, these grants are tied grants and have different obligations and rules for their utilisation. During this phase, in addition to these tied grants, the WB had initiated an innovative project titled "Gram Swaraj" in selected backward districts of Karnataka under which fixed amounts of untied funds were allocated to GPs over a period of five years to be utilised for implementing projects prioritised by rural citizens. So in the period between 2005 and 2010, GPs in Karnataka received a significant amount of resources for development programmes. As a result, Panchayat budgets increased significantly, doubling and tripling in many cases, and in some instances increasing nearly ten times.

From the above data, it is evident that the GP budget has seen a massive surge over the past ten years – the quantum of increase ranging from five times to as high as 15 times more in some GPs in 2000. Maranahalli, for example, had a budget of Rs 10,40,000 in 2000 but in 2010–11, it received Rs 17,425,933. The average quantum of increase across the GPs has been between eight–nine times their 2000–2001 budgets.

The business of democracy

The increased contestation that we are witnessing in Karnataka today can be seen as a direct response to increased GP budgets. A temporal analysis of the budgets and the total number of candidates contesting the three elections points in 2005, 2010 and 2015 show a sharp spike both in budgets and increased contestation in the 2010 election. Given the pervasive backwardness and the consequent lack of economic opportunities in this region, it is not surprising that GPs with increased budgets are seen as potential sites of economic opportunity. This is also visible in the way GP contestants have invested money in trying to woo voters. Candidates are reporting increased investments in elections made by them. Panchayats are seen as a career or business opportunity, especially by the parvenu. In this context, persistent traditional patronage networks continue to reproduce by adapting to the changing contexts.

Election expenses range from self-funded small amounts to large loans and range between 1000 rupees to more than 200,000 rupees. The majority of candidates spend between 30,000–45,000 rupees. The average expenditure across the 10 GPs was around 42,000 rupees in the 2010 Gram Panchayat elections.

However, the pursuit of electoral power also extracts high costs with some candidates losing all their savings in election campaigns.

In a context of persistent backwardness and poor economic opportunities, we find two distinct patterns emerging in these villages – a) those citizens, particularly young educated youth that view Panchayat politics as careers to engage in, and b) those that view this as a business opportunity to invest in. Developments in ICTs have opened new avenues of accessing information for those that were hitherto excluded from the game, thus levelling the playing field. GP is seen as a business opportunity in the villages; hence investments and payoffs or recovery of the investment with profits have become a common practice in the villages. A number of our respondents (those who were contesting elections for the first time) told us that they were not interested in continuing their political career once their term came to an end. On the other hand, those that perceived this as a career had started contesting SDMC elections as a prelude to the Panchayat elections.

However, although these changes have impacted the traditional patronage networks, they have not been completely displaced from this new game of electoral politics. Traditional power structures constantly try to reproduce themselves even now (Piliavsky, 2014). We find in our study of villages, that the traditional leaders have had to modify their game to accommodate these new entrants. In GPs where the attempt to put forth consensus candidates by the traditional leaders was not successful, in order not to lose face in the village, these leaders have started withdrawing from local politics. Instead, these leaders have shifted their political aspirations to the next level and are now concentrating on contesting in higher tiers of governance where political party backing brings an added layer of support in anchoring and reproducing patronage politics.

However, these pushbacks and rebellions by caste groups do not go unpunished. Given the region's violent history and strong embeddedness of patriarchy, the close proximity between the electorate and the elected representatives at times has led to immediate and harsh punishments for those that challenge the traditional power structures or demand accountability. A prominent SC leader from one of our sample GPs was murdered during the course of our study. In another GP, the female president was ousted through a no confidence motion after she objected to signing and authorising further funds for the construction of a pipeline that had remained incomplete.

The violence is not directed just at villagers but also at Panchayat functionaries. We came across a number of incidences where the then newly appointed Panchayat Development Officers (PDOs) were harassed, beaten and driven to committing suicide for not being complicit in efforts to misuse Panchayat funds.[12,13]

However, the situation is not as bleak as it appears to be. What is heartening is that such stories of corruption and violence are often juxtaposed with stories of rural citizens challenging corruption in Panchayats through the filing of RTIs and complaints to higher authorities and so on. Twenty-five years of decentralisation in itself has had an impact where excessive extraction of public funds is harshly punished by delivering electoral defeats. The increased devolution of funds has not only led to increased contestation by various caste groups, but also led to increased people's participation in the process of local democracy. Access to information about increased budgets has seen a rise in claims made by the citizens through increased participation in Gram Sabhas and Panchayat affairs.

A dualistic, intertwined model of democracy

What we are witnessing now in northern Karnataka is a flux where different actors are responding differently to increased devolution of funds in a region that has very few economic opportunities. On the one hand we are witnessing new caste alliances emerging as a counter to the traditional power structures in accessing local political space and on the other, we are also witnessing a better engaged electorate using formal tools such as RTI and weapons of "rude accountability" (Hossain, 2009) such as strikes or locking up of Panchayat offices in demand for better governance and accountability. These new forms of contestations and alliances are not only breaking down the caste-based traditional structure of dominance and patronage networks but also changing the rules of the game, thus forcing the traditional political elites to adapt and accommodate. Srinivas had predicted the potential power of voting on strengthening "caste consciousness" (Srinivas, 2002, p. 448). This continues to hold true nearly 60 years later. The role/importance of caste in politics has not become diluted or redundant in the changing local context. *Rather it has assumed new forms with identities now becoming crystallised around sub-caste categories.*

Interestingly, these shifts are happening within the larger context of deepening democracy where 25 years of decentralisation has institutionalised these democratic practices and institutions in Karnataka. Consequently what is emerging in northern Karnataka is a dualistic, contradictory but intertwined model of local democracy – one where the traditional political players use the local political space to regroup and remobilise in order to maintain caste cohesion, juxtaposed with the new entrants, particularly lower caste groups and educated youth, who are crafting new forms of inter- and intra-caste alliances to stake economic and political claims in a region where other economic opportunities are limited.

Notes

1 This chapter is based on a larger project on deepening democracy in Karnataka in collaboration with Dr Vijayendra Rao, The World Bank.

2 www.dcmsme.gov.in/schemes/pcgbk01x.htm
3 http://karsec.gov.in/prev_elec-fs.html, accessed January 05, 2018.
4 As reported in "Gram Panchayat poll results: various hues of democracy on display!", *Daijiworld*, May 17, 2010.
5 As reported in "Karnataka GP Polls: Results out, who won the poll?", *Deccan Chronicle*, June 6, 2015.
6 According to the 2011 Census.
7 As reported in February and March 2010.
8 As reported in Maranahalli in April 2010.
9 As reported in Aalanahalli in January 2010.
10 The Dalapati is a traditional title given during colonial times to individuals assigned to maintain local law and order. These were accommodated by the state under the Karnataka Village Defence Parties Act, 1964, He is considered to be on par with a police officer.
11 As reported in Kotehalli, April 2010.
12 As reported in "PDO's death highlights friction between GP members, officials", *The Hindu*, November 17, 2011.
13 As reported in "PDO's suicide: two suspended", *The Hindu*, November 12, 2011.

References

Ananthpur, K. (2007). Dynamics of local governance in Karnataka. *Economic and Political Weekly*, *42*(8), 667–673.

Ananthpur, K., Malik, K., & Rao, V. (2014). *The anatomy of failure: An ethnography of a randomized trial to deepen democracy in rural India* (Policy Research Working Paper No. 6958). New Delhi: The World Bank.

Ananthpur, K., & Moore, M. (2010). Ambiguous institutions: Traditional governance and local democracy in rural south India. *The Journal of Development Studies*, *46*(4), 603–623.

Banerjee, A., & Iyer, L. (2005). History, institutions, and economic performance: The legacy of colonial land tenure systems in India. *The American Economic Review*, *95*(4), 1190–1213.

Besley, T., Pande, R., & Rao, V. (2007). Political economy of panchayats in South India. *Economic and Political Weekly*, *42*(8), 661–666.

Besley, T., Pande, R., & Rao, V. (2011). Just rewards? Local politics and public resource allocation in South India. *The World Bank Economic Review*, 191–216.

Breeding, M. (2007). Vote-buying: Is it at threat to democratic policy representation? *PS: Political Science & Politics*, *40*(4), 821–821.

Chandra, K. (2007). *Why ethnic parties succeed: Patronage and ethnic head counts in India*. Cambridge: Cambridge University Press.

Chattopadhyay, R., & Duflo, E. (2004a). Impact of reservation in Panchayati Raj: Evidence from a nationwide randomised experiment. *Economic and Political Weekly*, *39*(9), 979–986.

Chattopadhyay, R., & Duflo, E. (2004b). Women as policy makers: Evidence from a randomized policy experiment in India. *Econometrica*, *72*(5), 1409–1443.

Chhibber, P. K. (1999). *Democracy without associations: Transformation of the party system and social cleavages in India*. Ann Arbor: University of Michigan Press.

Duflo, E. (2005). Why political reservations? *Journal of the European Economic Association*, *3*(2–3), 668–678.

Dunning, T., & Nilekani, J. (2013). Ethnic quotas and political mobilization: Caste, parties, and distribution in Indian village councils. *American Political Science Review*, *107*(1), 35–56.

Govt. of Karnataka. (2002). *High power committee for redressal of regional imbalances* (Chairman Prof. D M Nanjundappa). Govt. of Karnataka.

Govt. of Karnataka. (2005). *Karnataka Human Development Report (2005)*. Planning and Statistics Department, Govt. of Karnataka.

Govt. of Karnataka. (2015). *Village and Gram Panchayat Human Development Index (2015)*. Govt. of Karnataka.

Hossain, N. (2009). Rude accountability in the unreformed state: Informal pressures on frontline bureaucrats in Bangladesh. *IDS Working Papers*, *2009*(319), 1–35.

Inbanathan, A. (2003). Fixers, patronage, 'fixing', and local governance in Karnataka. *Sociological Bulletin*, *52*(2), 164–186.

Inbanathan, A., & Sivanna, N. (2010). *Scheduled castes, legitimacy, and local governance: Continuing social exclusion in Panchayats*. Bengaluru: Institute for Social and Economic Change.

Jaffrelot, C. (2003). *India's silent revolution: The rise of the lower castes in North India*. London: C Hurst & Co. (Publishers) Ltd.

Jaffrelot, C., & Verniers, G. (2009). India's 2009 elections: The resilience of regionalism and ethnicity. *South Asia Multidisciplinary Academic Journal*, (3).

Joshi, S., Kochhar, N., & Rao, V. (2017). *Are caste categories misleading? The relationship between gender and jati in three Indian states* (No. 132). Helsinki: World Institute for Development Economic Research (UNU-WIDER).

Kothari, R. (1970). *Politics in India*. New Delhi: Orient Longman.

Krishna, A., & Ananthpur, K. (2013). Globalisation, distance and disease: Spatial health disparities in rural India. *Millennial Asia*, *4*(1), 3–25.

Manor, J. (2007). Change in Karnataka over the last generation: Villages and the wider context. *Economic and Political Weekly*, 653–660.

Mansuri, G., & Rao, V. (2012). *Localizing development: Does participation work?* Washington, DC: The World Bank.

Nayak, P. (1995). Economic development and social exclusion in India. In A. De Haan & P. Nayak (Eds.), *Social exclusion and South Asia*. Geneva: International Labour Organisation.

Omvedt, G. (2006). *Dalit visions: The anti-caste movement and the construction of an Indian identity*. Hyderabad: Orient Blackswan.

Osborne, E. (2001). Culture, development, and government: Reservations in India. *Economic Development and Cultural Change*, *49*(3), 659–685.

Palaniswamy, N., & Krishnan, N. (2008). *Local politics, political institutions, and public resource allocation*. Washington, DC: International Food Policy Research Institute.

Palaniswamy, N., & Krishnan, N. (2012). Local politics, political institutions, and public resource allocation. *Economic Development and Cultural Change*, *60*(3), 449–473.

Piliavsky, A. (Ed.). (2014). *Patronage as politics in South Asia*. Cambridge: Cambridge University Press.

Rao, M. G., Raghunandan, T. R., Gupta, M., Datta, P., Jena, P. R., & Amarnath, H. K. (2011). *Fiscal decentralisation to rural local governments in India: Selected issues and reform options*. New Delhi: National Institute of Public Finance and Policy.

Srinivas, M. N. (1955). *India's villages*. Bombay, Calcutta, New Delhi, Madras, Lucknow, London, and New York: Asia Publishing House.

Srinivas, M. N. (2002). *Collected essays*. Delhi: Oxford University Press.

Srinivas, M. N., & Shah, A. M. (2007). *The grassroots of democracy: Field studies of Indian elections*. Delhi: Permanent Black.

Part II

CASTE AND MOBILITY

Views from below

5

M. N. SRINIVAS AND HIS "FIELD VIEW" OF SOCIETY

Some critical reflections

Joseph Tharamangalam

Introduction

Mysore Narasimhachar (M. N.) Srinivas is widely regarded as the founder of a new post-colonial Indian Sociology and Anthropology, the founder also of three important sociology departments in Baroda, Delhi and Bangalore, and the patron and Guru *par excellence* of more than one generation of Indian sociologists and anthropologists. Indeed, Srinivas effected a paradigm change in Indian Sociology in two respects. First, rejecting what he called the "book view" of society as promoted by Indologists and orientalists (thus anticipating the later critique of these Euro-centric and colonial approaches), he pioneered and promoted a "field view", one that was to be obtained by intensive field work using the method of "participant observation" in local communities, especially in India's "villages". Second, he merged the two different yet similar social sciences, sociology and anthropology, in an attempt to transcend what he saw as a colonial legacy that had no relevance for India. The former was developed by western scholars to study "their" own societies in the context of "modernity", and the latter to study "other" societies (in effect, their colonies in the non-western world), which they saw as primitive, homogenous, undifferentiated, and devoid of history. But perhaps ironically, Srinivas adopted the approach and the ethnographic method of British Social Anthropology as the most suitable for the study of Indian society. He did not reject historical and macro studies, but insisted that the insights obtained from intense field work should form the basis for the former kind of inquiry if it was to avoid the pitfalls of the "book view" and the use of what he termed "conjectural history".

This chapter attempts to critically examine the contributions as well as the shortcomings of Srinivas's Sociology and its method and their impact on subsequent Indian Sociology. While its main focus is Srinivas's "field view", it also briefly delves into the impact of the structural-functionalist perspective of British Social Anthropology that became intertwined with his field view.[1] The chapter is presented in three parts. Part II that follows this Introduction will discuss the field view in some detail, linking it also to the structural-functionalist perspective, briefly exploring both the positive contributions and the limitations of this approach. Part III, the concluding part, will suggest that despite Srinivas's many positive contributions and useful concepts like *Sanskritisation* and *Dominant Caste*, the micro-empirical village studies he pioneered may have led to the marginalisation of sociology (Deshpande, 2003) among Indian social sciences and its inability to deal with (even notice) some of the momentous societal changes that followed. Some notable examples of these are: what some observers have called *the rise of the plebeians*, the political assertion of the lower castes (Jaffrelot & Kumar, 2009), the impact of neoliberal globalisation and the consequent uneven and non-inclusive economic growth, and finally, perhaps most important, the rise of *Hindutva*, the Hindu nationalist movement that moved from its peripheral space in Indian politics and ideology in the early post-colonial period to the current dominant space and is now challenging what was considered the "idea of India" as a secular and democratic country that lasted for some four decades after India's Independence. It will suggest that while Srinivas's field view highlighted adaptation and resilience of the system – and interdependence, reciprocity and harmony – it could not, (and cannot) deal adequately with inherent conflicts and contestations in the system that would become more open in the context of macro-level electoral politics. Equally, Srinivas's field view also failed to adequately grasp the perspectives of the subaltern castes since it assumed a certain consensus about the ideological apparatus and norms that legitimised the system.

Interrogating Srinivas's field view: a bird's eye view without worm's eye view?

As mentioned, Srinivas anticipated by several decades the critique of the indological and orientalist view of Indian society, that he termed the "book view", basically an essentialised and static view of Indian society as a rigid and inflexible, *chaturvarna* system, governed by unchanging religious beliefs and laws, a view that had been adopted even by such western scholars as Henry Maine and Karl Marx. He challenged the validity of studying Indian society (or any society) through classical and religious texts (often representing the perspectives of the dominant caste or group) without observing how people actually live in contemporary India. "Manu" (India's classical lawgiver and ideologue), he said, "would be a bad guide for the field worker. Urban and upper-caste sociologists in India need to keep this constantly in mind" (1966).[2] In contrast to this, the "field view" he promoted would be based on intensive field work, using the method of participant observation.

It should be noted that this method had been developed by European (especially British) anthropologists to study "other societies" in the colonised world, tribal societies seen as homogenous and undifferentiated. There was no possibility of a "book view" here since these societies generally did not possess books or written sources. Srinivas first applied this method to the Coorgs of south India (Srinivas, 1952), a quasi-tribal society in transition, but soon moved on to apply this method to the study of the highly stratified, multi-caste villages in India. His students, colleagues, and other admirers followed suit; we are now inheritors of a large and rich body of literature on "India's villages" as well as on caste. The fact that Srinivas conducted his field work at a momentous historical juncture in a newly independent India that was embarking on what he thought would be a course of radical social transformation also influenced his work. He saw only a small window of opportunity to capture the true picture (structure) of the Indian village before it would disappear forever. Given this objective, the structural-functional perspective of British Social Anthropology came to his aid in exploring how all parts of the system contributed to the maintenance of the whole.

Some notable contributions

We have already mentioned how Srinivas's new approach and method made significant contributions to our understanding of Indian society, often leading to revisions in the conventional views. Especially notable here is how his field view radically revised the conventional view of the Indian village. He sees it as consisting of not four unchanging *varnas*, but numerous *jatis*, whose rank in the hierarchical system is often contested (with the exception of the two at the two ends of the continuum), a system that is fluid, resilient and adaptive to changing social forces. He formulates new concepts such as those of Sankritisation and Dominant Caste to help explain the dynamics of the system. Sanskritisation as a process by which lower castes imitated the cultural practices of the upper castes in an attempt at social mobility was, of course, familiar to local people and other observers, including Ambedkar who had termed it "the infection of imitation".[3] But Srinivas coined the term and popularised it, especially among sociologists whose use of the concept enhanced our understanding of both village society and the caste system. This is not to suggest that there is nothing problematic about the concept, as later critics were to argue. Srinivas also used the concept to explain two important aspects of India's civilisational process: 1) the two-way process by which Indian society has linked the little traditions of local communities and the great tradition of Indian (Sanskritic) civilisation – each absorbing aspects of the other's social, cultural and religious traditions and practices (including specific deities), and 2) the manner in which Indian (caste) society integrated, if not absorbed, outside immigrant groups, even non-Hindu religions.[4] Although, at a later stage in his life, he dared to write an "obituary" of caste (Srinivas, 2003), a premature one, as it turns out, it is worth noting that the conceptual framework he developed did emphasise the exceptional resilience of caste and the way it was able to adapt to changing conditions, even anticipating the way in which caste would occupy centre-stage in India's electoral politics (see especially Srinivas, 1987a, 1987b).

Srinivas made a significant contribution in critiquing the "myth of the self-sufficient village republic". In this instance, he drew on his field work as well as on historical evidence. He acknowledged and noted the link between the village and the broader economic and political forces, highlighting the role of the state.

As mentioned earlier, Srinivas also challenged the book view that saw caste as a system governed by religious ideologies and rules, and caste ranking determined by the idea of purity/pollution (previously challenged only by the rare Marxists). He highlighted the intricate and critical link between the mode of production and caste, between caste ranking and economic and political power. It is worth quoting his words here (Srinivas, 1987a, p. 525):

> A crucial feature of caste is its intimate linkage with production. Or rather, the process of production in traditional India cannot be disengaged from the caste system as it operates at the village or other local level. This gave the local caste system a measure of autonomy, insulating it from the consequences of attacks on the ideology of caste at the higher levels. For instance, the Bhakti sects in the early middle ages that led widespread anti-caste movements, but over the centuries the sects either became castes or reproduced caste-like hierarchies among themselves. *They do not seem to have taken note of the fact that as long as the mode of production remained unaltered, mere ideological attacks on the system would not change it.* (emphasis added)

This last point is worth highlighting in view of the fact that caste has now become more entrenched (far from annihilated) despite various policies designed to undermine it. Given this "intimate linkage", why did our founding fathers not seriously consider distributing land to the Harijans in the pursuit of reform and "annihilation" of caste? Even Ambedkar did little on this front, though he had fought for this in his Bombay days. Did they too, like the Bhakti sects, fail to take note of the significance of the mode of production in sustaining the system? Finally, why did Indian sociologists not raise such a significant policy issue? Interestingly, in a paper Srinivas published in 1987 titled "On Living in a Revolution" (1987b, p. 331) he notes (correctly, in our view) that India's generally upper caste policy makers have preferred the policies of reservation rather than land redistribution or public provisioning for basic social security and poverty reduction. He was commenting on the rising role of caste in electoral politics.

Deficits

Yet today, we are faced with the question, raised in the introduction to this chapter, of why Indian Sociology in general and Srinivas's Sociology in particular, is considered marginal in the social sciences grappling with significant socio-political changes and policy issues, including the new forms of assertion by the lower castes. I offer a few points that may help in answering this complex question.

Problematising the "field" and the sociologist's vantage point

A constant refrain in Srinivas's writings (and in his MA class in my days in the 1960s) was the statement: "Micro studies yield insights while macro studies yield perspectives; and the movement from one to the other is essential" (Srinivas, 1966). While this appears so self-evident, there are problems about locating the field and defining its boundaries, and about the sequence of the movement between the micro and macro studies he recommends. His own field work was focused almost entirely on the former with little reference to the latter. It is true that later in life, when he was also acting in the role of a public intellectual, he did bring a macro perspective to examine some issues, but neither he nor his students and followers actually did any significant research at the macro level, nor promoted any suitable method(s) to do this. The result was that the micro studies focused on the village social system with little reference to the larger social and political context. To take just one example, while untouchability was still practiced in Rampura during Srinivas's field work, independent India's law makers were engaged in a vigorous effort to enact effective legislation to abolish it and make it illegal. Untouchability was made illegal in January 1950. And Srinivas continued to visit Rampura for several years after untouchabilty had been illegal. Yet, he tells us little about how the villagers (both the perpetrators and the victims) were responding to it or if they were even aware of this except a few references to violent incidents. Today, the role played by caste in electoral politics, especially the formation of coalitions, can hardly be studied at the local level; the concept of the "dominant caste" itself is problematic in this context since different villages and regions often have different dominant castes.[5]

Related to the above is a second, and even more important question: is the highly stratified Indian village, physically segregated by caste (the touchables and untouchables divided by a highway between them in Rampura) one field or many fields, and in such a village where is the sociologist to perch to view the field, and from whose perspective? This is a critical question from Srinivas's own stated project of understanding the social world from the perspective of the people, which requires empathy and sensitivity. Yet, he settled in the *bullock house*, helped and guided in his field work by the "three important men" including the headman, all from the dominant caste. To be fair, let us note that Srinivas was acutely aware of his predicament, and did attempt to exercise a measure of freedom to visit and engage with untouchables and Muslims without offending his friends whose "common sense" included viewing prevailing caste rules and prohibitions as normal and natural (also as religiously sanctioned). He even tried various "tricks" (e.g., feigning ignorance) to achieve this goal, but with limited success. He has openly acknowledged these limitations in many of his writings. For example, this is what he says in his classic, *The Remembered Village:*

> My shortcomings as a field-worker are brought home to me poignantly when I contemplate the Harijans[6] and Muslims. I realize only too clearly that mine was a high caste view of village society. I stayed in a high caste area, and my friends and companions were all Peasants or Lingayats.
>
> (Srinivas, 1976, pp. 197–198)

But in the end, we have detailed descriptions of how his patrons and friends around the "bullock house" construct and see their social world, including their place in the village and their relationship with the low castes, but we have only some glimpses of what the world looked like from the perspective of the Dalits. We do have a few radical sociologists and anthropologists like Joan Mencher and Kathleen Gough who tried to see the world from the bottom up and we get from them a very different view. Mencher, who chose deliberately to spend more time among the Harijans, writes (1975):

> most of the Harijans I got to know tended to describe their relations with higher-caste people in terms of power, both economic (in terms of who employed whom, or their dependence on the landed for employment) and political (in terms of authority and the ability to punish). For Harijans both old and young, the exploitative aspect of hierarchy was what seemed most relevant, not the "to each his own" aspect. I did not once in my time in any of these villages hear from Harijans the usual rationalisations of inequalities.

Some 60 years after Srinivas's field work in Rampura, a research team (Wouters & Subba, 2010) visited the village and interviewed some elderly Harijans (including Ramanna who was a young man, and a *jita* servant[7] in the 1940s). They provide a different story – of extreme hard work under highly exploitative conditions, of constant humiliation and degradation (small portions of old, leftover food, etc.). They tell us that "the life-histories of elderly Harijans made us wonder whether the phrase 'cooperative relations' was not a euphemism for the practice of the higher castes in enforcing docility upon Harijans".

However, the keen eyes of Srinivas, the master field worker, did not miss instances of violence in Rampura, and he has documented these in many of his writings. For example, he describes an incidence of violence against Harijans in 1948 in the wake of a landmark legislation passed by the Mysore government that opened Hindu temples to the Harijans, and came into effect on the 13th day after Mahatma Gandhi's funeral. Interestingly, Srinivas notes:

> Harijans were persuaded, if not forced by younger caste Hindu leaders to enter their temples near Rampura. This may have led the Harijan leaders to think that at least the younger caste Hindus were keen on abolishing untouchability. If so, they were in for a disappointment. For, when some time later, Harijans in a few villages stopped performing certain festival duties they regarded as degrading, they were subject to economic boycott and beaten up, and their huts burnt.

Srinivas adds that the attacks were led by young men who had earlier appeared to espouse progressive causes (Srinivas, 1976, p. 200).

In a 1987 paper titled "The Caste System and its future" (1987a) he redefines the concept of dominant caste by adding two new criteria, the first of which is the "readiness to use violence to promote *collective ends*" (emphasis added). This statement is significant in that this is where he comes close to theorising violence, potential and actual, as functional in the maintenance of the system. Notable here is the phrase "collective ends" indicating his functionalist (and ideological) bias. Even as he criticised Dumont who "is extremely uncomfortable with the fact that in actual *jati* ranking, power is assigned a higher place than it ought to be on the ground of relative purity", he does not see a structure of power, backed by violence (systemic, structural violence) underlying the system, but only reciprocity and interdependence, even loyalty and friendship. We are left to conclude that his theoretical assumptions, and also his top-down view, prevent him from seeing the violence he documents as systemic and structural, and this despite his statement that such violence was exercised to promote "collective ends".

There is yet another dimension to this question, one raised by Dumont who argued that the hierarchical Indian village is not a community. Even as Srinivas examines the Indian village from a macro perspective in a later paper ("The Indian village: Myth and reality" in his Collected Essays, 2002) where he provides a historically informed, and more nuanced examination of the Indian village and its links with the larger society, he insists that the village is a "community" held together by reciprocity, cooperation and inter-dependence (even loyalty and friendship), arguing that a community can be hierarchical (as is also the case in Europe). While this may be a good, *ad hominem* response to Dumont, it still leaves us with some unease, and raises the question of why Srinivas does not explore the issue of violence more deeply or theorise it.

Impact of the structural-functionalist perspective

Srinivas's field view (as also those of his subsequent followers) becomes severely constrained by the theoretical assumptions he brings from the structural-functionalism of British Social Anthropology. It can be said that coming to the field with this perspective, he carried a bias prior to the investigation, constraining him to see Rampura (both as a village society and a caste society), as a place of reciprocity and interdependence, the only divisions and conflicts he examined resulting from "factions" led by different members of the dominant caste.

Finally, I wish to draw on my own research on Human Development (HD) in India to raise a critical question about which Indian sociologists have generally remained agnostic. This is the question about the contradiction between the country's high economic growth (and increasing prosperity) in tandem with exceptionally low Human Development indicators and exceptionally high rates of poverty, hunger, and malnutrition.[8] This is the contradiction between India's high economic growth and its exceptionally low human development. I have explored the issue in some detail elsewhere (most recently in Tharamangalam, 2017), and will state it here as briefly as possible. In all the measures used to gauge poverty, hunger, malnutrition, and HD by major international agencies, India has had an exceptionally poor record which places it below practically every country in east Asia, and even a few countries in South Asia in some respects. The Global Hunger Index has consistently reported over the past few years that India is home to the single largest pool of hungry people in the world (data also consistent with the Multi-Dimensional Poverty Index produced by IFPRI). As of 2016, India's rank in the UNDP's Human Development Index (HDI) stood at 131 out of 188 countries (one rank down from the previous year), below practically all countries in East Asia and 40 ranks below China's 90. And this is during a period when India has had high economic growth and increasing prosperity, significant advances in higher education (especially technical and scientific), spawning the now famous Indian middle class (also, arguably the largest in the world). This contradiction has remained a puzzle, almost defying explanation; neoliberal economists such as Bhagwati and Panagaria and spokespersons for the current government have engaged in denial,[9] and development economists like Drèze and Sen (2013) have explained this in terms of the deficits in India's development model, and its failures in public provisioning for basic social security. While I agree with the analysis provided by Dreze and Sen, I have also suggested that there is another dimension to the issue that has to do with caste. Very briefly, the argument is this. When we look closely at the data, we see the nature of the large pocket of chronic poverty and endemic hunger as consisting disproportionately of the historically excluded groups, especially the Dalits and Adivasis (indigenous people or literally original inhabitants) who together make up at least one-fourth of India's population.[10] In recent years, especially during the period of neoliberal globalisation, this historical exclusion has become more entrenched, widening the gap between the SCs and STs on the one hand, and the upper castes on the other. The question arises why Indian sociologists (except perhaps a few activist Dalit sociologists) whose special focus has been the study of the caste system, have neither seen this phenomenon nor said anything about this. Once again, it seems plausible to argue that their insensitivity to such a critical issue is rooted in the legacy of their field view, their functionalist perspective and their belief in value-free sociology, a legacy that has rendered them ill-equipped to deal with such a critical issue.

Is there a case, perhaps, that a methodology[11] created to study "homogenous and undifferentiated primitive tribes" may not be the best suited to study the highly stratified Indian society, especially after the nationalist movement had mobilised Indian society at a macro, national level, at a historically unprecedented degree?

Concluding remarks

To summarise the discussion, Srinivas is rightly celebrated as the founder of a new post-colonial Sociology in India, his valuable legacy still relevant today. Yet, I agree with other critics that the time has come to re-think and re-assess this legacy. As discussed above, the deficits in his Sociology stem from both his "field view" and his perspective of "structural-functionalism", these two closely linked and influencing each other. Despite his acknowledged attempts to overcome this problem, his view has remained a top-down one, as he took pains to explain. He emphasised the role of adaptation, mobility (through Sanskritisation) and resilience of a system he saw as based on inter-dependence and harmony, but he underplayed the role of conflict and violence (and possibly also, everyday forms of resistance) in the system. While instances of conflict and violence did not escape the keen eyes of this master field worker, these play a marginal role in his theoretical framework or in his analysis.

As mentioned in the introductory section, India today is going through major societal changes, including unprecedented levels of social mobilisation and contestation on the basis of caste and religion. Even as caste is becoming more entrenched and playing an increasing role in politics, the castes at the bottom, especially the Harijans, are experiencing increasing exclusion and disabilities resulting from the effects of neoliberal, non-inclusive growth (Tharamangalam, 2017), and from the upper-caste-led Hindutva movement.[12] Why, then, have sociologists remained agnostic about such critical issues? To repeat the argument advanced earlier, a plausible explanation is that the tradition of Srinivas's field view, combined with the functionalist assumptions, kept sociologists ill-equipped to deal with such issues. Consequently, these macro-level movements and tendencies (some of them system-threatening) seem to have remained beyond the purview of sociological inquiry. The challenge for a new generation of Indian sociologists is how they will move forward, adopting other, perhaps multiple, methods, and exploring other theoretical perspectives. As a former student and friend of Srinivas, this author has no hesitation in stating that Srinivas would have been only happy to see his followers boldly face such a challenge.

Notes

1 In fact, both can be seen as two aspects of his method if we take into account the two meanings of the term "method": as the method or tool for data collection and as a methodology – a theoretical perspective (and its assumptions) that informs the observation and especially the analysis, in this case the structural-functionalism of British Social Anthropology.

2 Manu is believed to have prescribed and codified the rules governing inter-caste behaviour, many of these humiliating and dehumanising the untouchables. It may be noted that Ambedkar, in a daring act of defiance and resistance, publicly burned the *Manusmriti,* Manu's classical text. As is well known, Ambedkar is the architect of the Indian Constitution and its many provisions aimed at improving the status of the lower castes. Less well known is the fact that he is also recognised today as one of the most (the most, according to many) erudite scholars of the twentieth century (at least according to many experts who participated in the Ambedkar International Conference in Bangalore in 2017) with two PhDs in Economics and Law from the LSE and Columbia University respectively. He is also the author of some landmark critical works on caste including the now famous *Annihilation of Caste* (2016). It is noteworthy that towards the end of his life, Ambedkar converted to Buddhism, together with a few hundred thousand followers, after he came to the conclusion that untouchables would never escape caste within the fold of Hindu society.

3 See Ambedkar (2014), especially his seminal essay, *Annihilation of Caste.* I think it is relevant to note here that neither Srinivas nor his students and followers who focused so much attention on the caste system seem to have read Ambedkar's writings on caste though the latter had made some landmark contributions to the study of caste in which he also explored the relation between caste, patriarchy and demography (Kannabiran, 2009). Ambedkar's name was never mentioned in any discussion of caste in the MA classes in the Delhi Sociology department during the years I was a student there (1967 to 1970). It may be added that even Gandhi's name was rarely mentioned except in the context of "exploding the myth" of the *self-sufficient village republic* that Gandhi advocated as a model for India's future development.

4 It is interesting to note that some of these groups, such as the Parsis and the Syrian Christians of Kerala, were given relatively high status in the system though strictly speaking, they may have been outside the caste system. It would seem that the actual status they enjoyed depended on the specific historical context of their arrival and especially the economic role they played in the system (e.g., as wealthy traders).

5 Let me note that in a 1987 (1987b) essay titled "On living in a revolution", Srinivas provides an insightful analysis of the caste issue at the macro level. He notes many of the dilemmas and contradictions, including how untouchability continues to persist in spite all the legal and constitutional provisions. He also notes how upper caste politicians are reluctant to initiate or support anti-poverty programmes, preferring to promote "reservations" for increasing numbers of "Other Backward Castes", a strategy they see as more rewarding politically.

6 The term "Harijan" (people of God) was introduced by Gandhi to refer to the untouchables. Later, many untouchables rejected the term as patronising, preferring instead the term "Dalits" (literally "the broken ones"). Today, Dalit is the preferred term to refer to this group. I have chosen to use both terms depending on the context of its use.

7 A servant who enters into a contract with the master, generally for two–three years, and is paid a sum of money in advance (exclusive of food and clothing which also will be provided by the master). Most *jita* servants renew their contract before its expiry, borrowing another sum of money. In effect, it does not appear to be very different from bonded labour (Srinivas, 1967, pp. 27–28).

8 Drèze, Jean and A. Sen (2013) refer to this as India's contradictions. Sen has dealt with this extensively in many of his writings. Others have seen it as a puzzle. It may be viewed as a form of Indian exceptionalism.

9 Prime Minister Modi himself seemed to be clueless on the issue when during a visit to Kerala, he dared to compare Kerala's developmental model with that of Somalia, quoting some figures of high infant mortality rates among Kerala's Adivasis. He seemed to have been completely ignorant of the international debate about the "Kerala Model of Development" that has stood out for many years in India and the world for its high HD achievements (consistently ranking first among all Indian states) and reduction of poverty and malnutrition (lowest among Indian states). By contrast, Modi's brand of the "Gujarat model" has been considered a highly flawed one by HD experts, one that catered to a minority of the rich. For my own assessment of the Gujarat model, which I have called a quintessential example of *crony capitalism*, see Tharamangalam (2014). Modi's ill-informed remarks drew a voluble reaction in Kerala, widely reported in the media. See, for example, www.bbc.com/news/world-asia-india-36264001

10 Note that the category of Scheduled Castes (SCs) does not include all ex-untouchables or Dalits, but only those who are counted as Hindus. Those who converted to Islam and Christianity have been excluded from this category, a double exclusion since they are not entitled to the benefits of reservation provided to STs. Those who are Sikhs or Buddhists were also originally excluded, but were added to the list later. While SCs make up about 17 per cent of India's population according to the Census 2011, Dalits, regardless of religion, probably make up about one-quarter of the Indian population, if not more. If this is so, the two excluded groups of ex-untouchables and the indigenous population make up at least one-third of the total Indian population.

11 In both senses of the term, mentioned in note 1 above, as the tool for data collection and as a theoretical perspective that informs the observation and especially the analysis, in this case, the structural-functionalism of British Social Anthropology.

12 While the movement tries to co-opt the Dalits to consolidate the Hindu *Parivar* (family) and Hindu majoritarian dominance in the electoral system and the power structure, there is also increasing discrimination and violence in

the service of upper-caste-defined *Hindutva* causes such as "cow protection". Predictably, this is not entirely lost on Dalit politicians. Dalit leader and former chief minister of Uttar Pradesh Mayawati recently threatened that Dalits would follow the example of Ambedkar and convert to Buddhism if such discrimination and violence continued. See *The Deccan Chronicle*, December 11, 2017. Retrieved from www.deccanchronicle.com/nation/politics/251017/up-bsp-president-mayawati-threatens-to-convert-to-buddhism.html

References

Ambedkar, B. R. (2014). *Annihilation of Caste*. (Edited and annotated by S. Anand, and with an Introduction by A. Roy). Verso Beteille, Andre (Ed.) 1969. Social Inequality: Readings. New York: Penguin.

Deshpande, S. (2003). *Contemporary India: A sociological view*. Delhi: Penguin Books.

Drèze, J., & Sen, A. (2013). *An uncertain glory: India and its contradictions*. Princeton, NJ: Princeton University Press.

Jaffrelot, C., & Kumar, S. (Eds.) (2009). *Rise of the plebeians? The changing face of the Indian legislative assemblies*. New Delhi: Routledge.

Kannabiran, K. (2009). Sociology of caste and the crooked mirror: Recovering BR Ambedkar's legacy. *Economic and Political Weekly*, 35–39.

Mencher, J. (1975). Viewing hierarchy from the bottom up. In A. Beteiile & T. N. Madan (Eds.), *Encounter and experience: Personal accounts of fieldwork*. New Delhi: Vikas.

Srinivas, M. N. (1952). *Religion and society among the Coorgs of South India*. Bombay: Asia Publishing House.

Srinivas, M. N. (1966). *Social change in modern India*. Berkeley and Los Angeles: University of California Press.

Srinivas, M. N. (1967). The cohesive role of sanskritisation. In Masons (Ed.), *Unity and diversity: India and Ceylon*. London: Oxford University Press, pp. 221–235.

Srinivas, M. N. (1976). *The remembered village* (No. 26). Berkeley and Los Angeles: University of California Press.

Srinivas, M. N. (1987a). The caste system and its future. In P. Hockings (Ed.), *Dimensions of social life: Essays in honour of David Mandelbaum*. Berlin: De Gruyter Mouton.

Srinivas, M. N. (1987b, December). On living in a revolution. *Berichte*.

Srinivas, M. N. (1992). *On living in a revolution and other essays*. Oxford University Press.

Srinivas, M. N. (Ed.). (1996). *Caste: Its 20th century Avatar*. Delhi: Penguin.

Srinivas, M. N. (2002). The Indian village: Myth and reality. In *Collected essays*. New Delhi: Oxford University Press.

Srinivas, M. N. (2003). An obituary on caste as a system. *Economic and Political Weekly*, *38*(5), 455–459.

Tharamangalam, J. (2014). *Gujarat: A seriously flawed "model" of development and democracy, blatantly tilted towards interests of a small minority*. Counterview. Retrieved January 19, 2018, from https://counterview.org/2014/04/08/

gujarat-a-seriously-flawed-model-of-development-and-democracy-blatantly-tilted-towards-interests-of-a-small-minority

Tharamangalam, J. (2017, July 21–23). *India's paradox of "hunger amidst plenty" has a name: Caste-based discrimination and exclusion.* Paper presented at International Conference on Reclaiming Social Justice, Revisiting Ambedkar, Bengaluru, India (forthcoming in conference proceedings).

Wouters, J. J. P., & Subba, T. B. (2010). Revisiting Srinivas's 'remembered village'. *Journal of the Anthropological Society of Oxford, NRE Series*, 1–2, 44–65.

6

ONE VILLAGE AND MANY CHANGES[1]

The problematique of understanding mobility among Dalits

Sobin George

Introduction

Scholarship on caste has generated compelling discussions on its changes and continuities from various standpoints and contexts. It has discussed patterns of mobility (both social and economic), political mobilisations, assertions, citizenship rights and questions of development by invoking possibilities of change due to processes such as loosening of various forms of dependency of lower castes on upper castes, weakening of traditional institutions, political mobilisations and proliferation of the values of democracy, dignity and technology (Jodhka, 2002; Srinivas, 2003; Gupta, 2005; Shah, 2007; Manor, 2007), among others. Some of these studies have also offered persuasive explanations on the nature of caste relations based on dominance (Srinivas, 1994), differences (Gupta, 2000), conflict and accommodation (Manor, 2012). It is generally agreed that occupational rigidities started fluidifying (Mayer, 1996) and Dalits who benefited from reservation policies moved up both socially and economically (Ram, 1995; McMillan, 2005). While some of these studies highlighted the upward mobility of Dalits, others cautioned on the continuing barriers in their path of upward mobility. Deshpande (2011), for instance, noted that the Dalits who moved out of traditional occupations tend to get entangled at the lower rung of the economic class. The reasons are several, ranging from higher initial levels of vulnerabilities and discrimination (Madheswaran & Attewal, 2007) to lack of social capital and effective business network support and economic exclusion (Jodhka, 2010; Harriss-White & Basile, 2014). While some scholarship draws attention to the opportunities of the "new and caste-neutral occupations" available for Dalits (Kapur, Babu, & Prasad, 2014), another set of scholars argues that Dalits are "unfavourably included" (Teltumbde, 2008;

Thorat, Mahamallick, & Sadana, 2010) in such markets and the market principles and ideologies could not "break down India's caste-based social order, but reinforce it in some ways" (Harriss-White & Basile, 2014). Similarly, studies have also shown that although Dalits are politically assertive in many parts of the country, the local elites continue to be the power centres in local self-governments (Inbanathan & Sivanna, 2010; Martin, 2015).

These articulations and explanations of *change* have indeed helped us to comprehend caste relations with its context specificities. However, over a period of time, the scholarship has travelled in such a direction that these explanations started getting universal meaning and it has become quite *fashionable* to explain changes in such common directions, taking the "experiential dimensions" (Guru, 2016) of caste completely out of its situations. Caste explanations thus have tended to undermine an important epistemological position of *many experiences of caste* of Dalits. Arguably, this, to a certain extent, also has led to a kind of discomfited polarisation of caste literature in which, broadly, one group supports mostly the thesis of fluidification of caste and the other firmly believes that nothing has significantly changed in the original content of caste, although its forms might have changed. The question that this chapter puts forward therefore is: what is the fundamental problem in explaining the changes in caste relations in a uniform fashion by borrowing from or forming convenient categories when there is no universal experience of caste? The chapter attempts to address this question by drawing from ethnographic field work conducted in a village named Meenkera in Bidar district of north Karnataka. It presents the narratives of Dalits from Holeya, Madiga and Waddar sub-castes in the village to illustrate the possibilities of *many experiences* of change and continuities of caste relations in various spheres of their life. It must be noted that their views come from a reality that they or others experienced in a particular context and situation.

Castescape of the ethnographic site

One of the important considerations for locating a site to conduct the present field study was the trajectory of change in social, economic and political spheres for Dalits. Similar to other parts of the sub-continent, explanations on the changing nature of caste in Karnataka have also been mainly around questions of hierarchy, dependence, dominance, differences, change and continuity (Karanth, 1996; Charsley, 1998; Armstrong, 1998). The major change that these studies highlighted is the decline of dependency, both economic and social, on agriculture and thereby on others by the lower caste groups, which arguably have re-arranged caste relations in rural Karnataka (Karanth & Ramaswami, 2005). The reasons for the diminished dependency on agriculture as these studies pointed out were the shift to smaller and marginal farming because of large landholdings getting smaller due to intergenerational transfer of land, widespread commercialisation of agriculture following the collapse of bonded/contractual relations, prevalence of payments in cash, mechanised capitalistic modes of organisation of agriculture production and finally dependence on resources and avenues outside the agricultural sector. At the same time, some of these studies also highlighted the continuity of patronage by the upper- and middle-level politically dominant caste groups on Dalits in different spheres in varying forms. For instance, Karanth (2004) noted that the continuing cultural subordination of Dalits to Vokkaligas in southern Karnataka led to their dependence on them in availing the benefits of state welfare schemes. Political mobilisations and processes from below that have changed the caste-based voting pattern and the emergence of caste organisations with their "considerable influence" in electing governments and state interventions through targeted programmes for Dalits were also major forces of change in rural Karnataka (Manor, 2007).

The village Meenkera, where the field study was conducted, is situated in Humnabad taluk of Bidar district in north Karnataka region. The village is nearly 3 kilometres away from the Hyderabad–Mumbai highway and is 58 kilometres away from Bidar district headquarters. As per the 2011 Census data, the Dalit population in Bidar is as high as 23.5 per cent, which is more than the state average. Dalit sub-castes in the district are Holeya, Madiga, Mahar, Mangs, Chambhar, Waddar, Dhor, Samagara, Mochi, Panchama, Dore and Chalvadi and Beda in the order of their share of population. Bidar district has a considerable presence of Dalit population in agriculture labour. The data shows that as much as 72 per cent of the Dalit households depend on casual labour in agriculture for a living (George, 2016). Bidar, however, also gives a different picture of change for the remaining Dalits, who are outside the agriculture sector. Interestingly, the share of self-employed Dalits in the non-agriculture sector in rural areas of Bidar is 18 per cent, which is higher than any other district in the state (George, 2016). The share of non-literates among Dalits in Bidar is less and the share of Dalits who have completed secondary education and above is more, as compared to some of the south Karnataka districts (George, 2016). The region also has a rich history of Dalit activism with the strong influence of Ambedkarite movements, which, possibly, have acted as a force of change from below (Assadi, 2004).

Meenkera village was compelling enough for us to undertake a study on the changes in caste relations because of several reasons. First, the village is numerically dominated by Dalit castes of Holeya, Madiga, Waddar and Bhadari together. As per the data from the village office of Meenkera, out of 909 households in the village, 331 belong to Dalits, 297 belong to OBC groups including Madiwal, Nekar, Badigar, Kuruba, Swami and Uppar from Hindu community and Sayed, Pathan, and Shaik from Islam. As many as 213 households belong to ST groups which include Tokre Koli, Kadu Kuruba, Gond and Kapadia. Only 68 households in the village belonged to the upper- and middle-level caste groups, which include various sub-groups of Lingayats,

Gowdas and Brahmins. The village has 754 Hindu, 113 Muslim, 30 Buddhist and 12 Christian households. Second, as it is true for several parts of rural India, caste relations have rearranged themselves in Meenkera with the movement of landed Brahmins and rich Lingayats to cities. The land holding patterns have changed considerably with the relocation of landed caste groups along with its lease or sale as well as the shrinking of farm land due to intergenerational transfers. The economic mobility of some of the Dalits, especially those who are from Madiga and Holeya castes, have helped them buy lands which are sold by Brahmins and Lingayats. For instance, we have seen some Madiga and Holeya households which possessed nearly 10 acres of agriculture (irrigated) land in Meenkera. While nearly 24 per cent of Dalits in Meenkera are landless, a majority of the Dalits in the village possessed land. This included both agricultural and non-cultivable land. It should be also noted that non-agricultural land has real estate value in the villages as it is near the Mumbai–Hyderabad national highway.

Third, changes have also appeared in the economic relations of the village with the availability of non-agriculture occupations and changes in contractual agriculture relationships. The villagers are mostly small farmers and agricultural labourers. However, most of them, especially those from the Dalit castes, also take part in non-agricultural activities since the opportunities available in the agriculture sector are less. Each household has at least one member in non-agricultural employment. While the elders in the family are mostly tied to agricultural labour, the second generation tended to move out to other avenues like casual labour in non-agricultural occupations, own agriculture and other informal works in the private sector. The shift outside agriculture is complete in the third generation for many Dalit households. We could see Dalit families sending their children to school, if possible to English medium schools. Also, we could seldom find any young Dalit wishing to continue in her/his traditional occupation.

While economic relations, to an extent, have changed, the spatial organisation of the village along the lines of caste remains as it is. For instance, Lingayat, Gowda, Kuruba, Muslim and Brahmin families live on one side of the main road with separate lanes, whereas the lower caste Hindus, including Holeyas, Madigas, Waddars and STs including Kabaikas and Kadukurubas live on the other side of the road, but in separate lanes. Streets in this part of the village are also named after caste names. The settlement of upper caste people is called *Uru* (main village) whereas Dalit settlements are named Harijan galli, SC galli, Madigara galli, and Waddar galli. *Uru* is also further classified into Lingathara galli, Kurubara galli, Muslim galli, etc. The civic life of the people is visibly segregated. For instance, the village has two Anganwadis: one for Dalits, STs and Muslims and another for caste Hindus. There were also separate places for worship (temples) for each caste group. As Dalits are generally not expected inside the common temples of the village, they worship "their own gods" in their locality.

Many narratives

The following sections present the narratives of Dalits from various individual castes and walks of lives on the changes that they experienced and perceived in different spheres of life. Although it is not important to group these narratives across various available explanations on caste relations for the epistemological position that the present chapter takes, these are broadly organised across some frequently emerging themes for the ease of presentation. It must be noted that these themes and the group of respondents are not mutually exclusive categories and the narratives reflect only their experience of caste in a particular space or situation. The real names of some of the respondents are not given in the chapter since they wished to remain anonymous.

Denial of caste practices and avoidance: the young, educated and the converted

Some of the Dalits in Meenkera vehemently denied the existence of caste-based discrimination against them in the village. They include those who are in public sector jobs and who work in private factories, students and some who have converted to Buddhism from Holeya caste and to Christianity from Madiga caste who are also economically better off among Dalits and vis-à-vis other caste groups. Their narratives were mostly about non-dependency on upper castes, which they think freed them from caste ties. They view caste and caste ties as something which has nothing to do with their present day life. For instance, 28-year-old Suhasini (name changed), an Anganwadi teacher from Madiga community, explains the change as follows:

> Earlier we (Dalits) were supposed to work in Gowda's (middle level caste) and Kulakarni's (Brahmin caste) house. Our livelihood came from that. Now we earn independently and are independent. Earlier, it (untouchability) was there. Earlier, in our village, we were supposed to wash our clothes after upper caste people had their turn. In the river, if they stand in the upper end, we used to stand in the lower end. While washing, we had to be very careful and water from our clothes should not touch them. If it touched them, they used to scold us and ask us to stand silently till they completed their work. They used to throw food into our hands. But now, things have changed. We only stand at the upper end and they stand in the lower end.

They were very emphatic in narrating the changes that had taken place in the highly rigid sphere of sharing common resources such as water and interdining. Maria, a Dalit woman who converted to Christianity, for instance, noted that they now not only share the common water resources available in the village, but control them as well. For them, the change is "absolute" and in some instances went even to the extent of reversal of existing caste practices. For instance, Maria noted that upper caste people used to wash their clothes at the upper end of the only water stream available in the village whereas the Dalits were expected to stand at the lower end. The concept of *purity* and *pollution* was so intense that upper caste women did not tolerate even the water drops from the clothes of lower castes falling on their clothes. However, now Dalit women have captured the space at the upper end of the water stream, which pushed the upper caste women to the lower end. Similarly, many of them also shared that young upper caste male members do not hesitate to eat in their houses, if invited, on special occasions. As some Dalit women noted, upper caste men also find their Dalit friends' houses to be places where they can consume alcohol and non-vegetarian food (including beef) secretly.

Although for some of them, like the Anganwadi teacher from Madiga caste, the young college student from Holeya caste and the young woman who converted to Christianity, caste practices by upper caste in the village have almost disappeared in social relations, this was not the case always for others from the same age and occupational group. Their narratives reflected the independence that they have achieved and the reduced interactions with upper caste people, which *empowered* them to avoid caste-based discrimination in the village. For instance, the educated and employed Dalits felt that it was the economic independence that changed their relations with upper castes, whereas the neo-Buddhist and Christians from this group think that their religious conversions also freed them from the ritual obligations with caste Hindus. While churches and priests from their own religious community took care of the religious rituals connected to birth, death, marriage and other functions of Christians, neo-Buddhists have their own priests to perform these. Although religious conversions have not changed the perceptions of upper caste Hindus on Holeyas and Madigas, the converts could isolate themselves from caste Hindus. Similarly, it was also found that they tend to avoid environments that might be favourable for discrimination. This was visible in the choice of non-agriculture occupations. They resist the caste practices by not being part of the caste centric agriculture and other traditional occupations. Similarly, they are aware that upper castes tend to discriminate against Dalits in certain private spheres and hence choose to avoid confrontations by keeping away from such private spheres, which mainly include their home, family/community and temples.

While their narratives reflected more of non-dependency, avoidance and complete change in caste relations in the village, the story was not the same in other spheres of life outside the village. While they enjoyed dignity and respect in their village, some of them noted that they experienced caste-based discrimination at the workplace including the public services. For instance, a police constable named Rudrappa (name changed) who belongs to Madiga caste noted that while he receives respect from all in the village, the experience was different at his place of work. He said he experienced different forms of discrimination at the workplace including "delaying promotions, assigning difficult and lower level jobs, delaying clearing of personnel files, delaying housing facilities and travel allowance and asking unwanted questions for reimbursing eligible expenditure bills", among others. To quote his words:

> They (upper caste superior officers) will assign good work (easy, non-humiliating, less cumbersome ones) to people from their own group. Though we face caste discrimination in our service, we are helpless. They are more organised and close to power. They can do anything including filing false charges against us if we revolt. What can we do? What strength do we have to oppose the discrimination? Low grade employees, who are mostly Dalits, like C grade and D Grade employees, can't do anything to higher officers. I must also say that I get respect from all caste groups in my village.

Continuities and confrontations: the young, educated, the alert and the articulate

Articulations of caste relations as more of assertions and confrontations emanate from a strong Dalit consciousness, which recognises that possibilities of unfavourable caste practices against them exist. They tend to challenge caste-based discrimination and look for changes through assertions. This kind of narrative also emerged from the educated young members, public servants, migrants and some neo-Buddhists, who are actively participating in political mobilisations in Meenkera. While they found that they could insulate themselves from caste-based discriminatory practices of upper castes in the village, they recognise the possibility of confronting caste in other spheres of life both inside and outside the village. Like in other parts of the country, growing Dalit assertions have led to different forms of confrontations between Dalits (mainly Holeyas and Madigas) and Lingayats and Gowdas in Meenkera. The reasons for such confrontations relate mostly to practices of untouchability and the experience of humiliations such as discriminatory practices in public places and public harassment of Dalit women by upper caste men. The nature of confrontations ranges from open heated arguments to physical violence.

The youngsters from the Dalit community, especially those who are educated and affiliated with Dalit activist groups such as Karnataka Dalita Kriya Vedike and Jai Bhim Dal, tend to question the discriminatory practices. They narrated several such instances of resisting public humiliation by the upper castes in the village. For instance, it had been a practice to make Dalit women dance in front of a crowd (of drunken upper- and middle-level caste men) in connection with the temple festivals of *Marigemma Jatre* in several north Karnataka villages including Meenkera. People never used to question this since it was considered as an inevitable part of the temple festival rituals. It was believed that such performances by Dalit women in connection with the festival would repel evil forces which otherwise might bring misfortunes such as drought, famine and epidemics to the village. Although such performances sometimes would end up in outrageous violence including rape on Dalit women, it was silently sanctioned in the name of religion and faith. Young Dalits in Meenkera, with the help of BSP activists and members of Dalit organisations such as Jai Bhim Dal and Samatha Sainik Dal opposed this practice in 2013 despite some differences of opinion even within Dalits on such practices. They not only resisted the performance but managed to get media coverage on this issue, which was eventually taken up by Bahujan Samaj Party (BSP) and other progressive movements. Despite strong resistance from all others including elders from Dalit castes, they were able to stop the practice in Meenkera. Another instance of resistance, which also led to physical violence, was around the celebration of the birthday of Dr B. R. Ambedkar on 14 April. As some of the respondents noted, the upper castes used to create problems while Dalits celebrated Ambedkar Jayanti in Meenkera. Once in Mannaekhally, a small town nearby, some upper castes defaced the statue of Ambedkar with shoe garlands, which led to conflicts. Dalits from nearby villages mobilised at Mannaekhally. The protest took violent forms, which even led to burning of buses, *hartals* and subsequent deployment of military force. Several Dalits and non-Dalits were injured and hospitalised. Premnath, a local leader of BSP and a Dalit activist, said such mobilisations increased the confidence of Dalits to challenge discrimination.

Some of the young and educated Dalits from Holeya and Madiga castes who are part of caste organisations have also taken an extreme position on caste, related to its unchanging nature at the core notwithstanding the changes in the periphery. Their articulations reflect that caste continues to be a strong social force that divides people despite several reformative and correctional initiatives from above and resistances from below. Certain narratives that we recorded in Meenkera highlight several covert and overt practices of caste in the public and private spheres in various forms. The practices of untouchability, as Dalit activists say, have not changed in the private spheres. Premnath, a young Dalit activist, narrated an incident which he experienced recently in the village:

> Dalits wrongly say that there is no caste discrimination here. Upper caste people don't allow us inside their houses. You believe me or not, there is a person by name Baburao Biradar (name changed) in

> our village, who belongs to the Lingayat caste. He invited us to take tea in his house. But he gave separate cups to us which were kept outside. I am speaking about my village. When we are outside, he will consume food with us in the same row. But he behaves differently in his home.

It is also important to note that the perceptions of change and continuity of caste practices varied within an individual caste or even family across different age and occupational groups depending on what they consider as "real change". The reference point of change and its relative dimensions are important here. While an old Dalit agricultural labourer considers that invitation to a Lingayat family or Lingayats coming to their locality or Dalits sharing common resources along with upper castes are "changes", it is possible that a young educated Dalit from the same caste or family, on the other hand, would consider the differential treatment they experience in these spheres as caste continuity. For them, the real change means equal rights and equal participation with dignity. Hence, when an old Dalit person says, "it (untouchability) was there earlier, but not now", what she/he normally implies is that the classical form of untouchability is not being practised explicitly. The family of Arjun, which belongs to Holeya caste, is an interesting example on the differing perception of caste practices and discrimination among family members. Arjun works as a supervisor in a small construction company in Bidar town. He lives in Bidar with his wife and two children. He has two elder brothers, who live in Meenkera along with their parents, wives and children. His brothers are agricultural labourers in the agriculture fields of a Lingayat family. They have been working for the same person for the last three years. As Arjun notes, they never bargain for wages and agree to what the landlord decides. He also found that the Lingayats still serve them food from a distance. They are also not allowed to eat in the same place together even in the agriculture field. Similarly, if they carry food, they are expected to keep it away from labourers of other castes in the field. While Arjun attributes these practices to discrimination, his brothers find it absolutely alright. Savita, a young Dalit woman from Holeya caste in Meenkera, explains how she and her father differ in their perception of the untouchability practices of the upper castes:

> When I was small, I used to go to their (Lingayats) houses with my father and mother, who were workers in their farm. What I can recall is that they used to pour tea from a distance into our cups, which they continue even now. I used to tell my father that if they pour tea from a distance we should not drink it. I always questioned my father for his submissive behaviour. My father, however, used to say that this is a years-old practice and we should be fine with it.

Changes with continuities: agriculture workers, small entrepreneurs and the elderly

While the young, educated, activists and economically better off sections of Dalits in Meenkera shared the experiences of complete denial of caste experience, avoidance, assertions and confrontations and its unchanging nature, some section of Dalits in the village who are mainly agriculture workers, small farmers and small entrepreneurs and elderly from Holeya, Madiga and Waddar castes shared another view of caste. They view change as a multi-layered process with changes and continuities of caste practices in different spheres. They agree that caste relations are changing; however, they recognise and accept the caste hierarchies which govern the social and economic relations in the village. Despite the advances in education, mobility and non-agricultural employment, they still hold on to the view that the economic dominance of Lingayats and Brahmins is a reality even now, and several Dalit families depend on them for a living. For instance, while work opportunities in agriculture decreased, the upper caste Brahmins, Lingayats and a small but dominant group of Gowdas continue to be in control of new employment opportunities in sectors such as stone cutting, clay-tile factories, co-operative societies, small shops, transport, schools, private dispensaries, public contract work, etc. in the village. Dalits perform lower level works such as daily wage labour in stone cutting and clay factories, or as cleaners of trucks and other transport vehicles, as cleaning staff in hospitals, as drainage cleaners and daily wage workers in public works and as peons in schools. Some of these are not only lower level employment in the work hierarchy, but exploitative as well. While some among them such as peons in schools and cleaners of vehicles are not employed in their caste-related occupation, others like drainage and toilet cleaners continue to be in their caste-related unclean occupations. In effect, as per their view, even though some of them could move out to the so-called modern occupations, it is incorrect to assume that caste does not get reproduced in these occupations, as Thorat et al. (2010) and Harriss-White and Basile (2014) have pointed out.

It is reported by these groups of respondents that only a few Dalit families that possess irrigated land and have members who have well-paying non-agricultural jobs could get their children educated and come out of the dependence on upper castes. As studies have already revealed, the practice of *Jiithagara*, which is a form of contract labour that evolved from the old bonded labour system in Karnataka wherein a lower caste person or family agrees to work in the land of the upper caste for a year for either mutually agreed lumpsum wages or as fixed by the landlord, has been mostly replaced by the non-contractual daily wage system. We could find only two Dalit families from Waddar caste continuing this practice due to their loyalty to the Lingayat families. The availability of other seasonal jobs in public works and non-agricultural contract works is reported to have helped Dalits to reduce their dependency on upper castes. However, as the elderly Dalit members noted, there are (moral) obligations prevailing in the villages between landed upper castes and poor Dalits. As they noted, since the available non-agricultural jobs are uncertain and limited, poor Dalit families that do not have irrigated agricultural land and salaried members have to seek work in the fields of Lingayats and Gowdas, continuing the patronage relationships. Though Dalits manage to get wages as per market rates, they appear to be loyal to the landlords for want of work. Some of them also noted that they "pretend to be loyal" to Lingayats and Gowdas for getting work.

The narrative of "change with continuity" was true for small farmers and small entrepreneurs, who moved out of agriculture labour and other caste occupations, as well. For instance, the small farmers (those who have less than 5 acres of land) from Holeya community, who could manage to buy land in Meenkera, continue to depend on upper caste merchants to sell their produce as the local market is controlled by the upper castes. They have to be in a "good" relationship with others in order to share the common resources such as water and make use of various agricultural support programmes of the state. Similarly, the small Dalit entrepreneurs in Meenkera realise that their survival is possible only through the business network controlled by the upper caste people. As Jodhka (2010) elaborated in his study on the Dalit entrepreneurs in Haryana and western Uttar Pradesh, the Dalits in Meenkera too do not possess the social capital and network essential for such businesses to either flourish or get the necessary support at the time of unexpected shocks.

Caste-based social relationships also showed changes in Meenkera, as several scholars have already found in other parts of the country. These villages witnessed changes in the practices of exchange of water and food. While it is true that there is no apparent prohibition on the exchange of food and water among various caste groups in the village, one could see restrictive practices in interdining in private and public spheres. It is very rare to see tea shops providing separate cups/glasses for tea nowadays in Meenkera. But the shopkeepers, like in many other parts of the country, manage it by using disposable cups for all. Nevertheless, there are some changes in interdining, especially during occasions such as weddings, child naming, post-funeral rituals and other family functions. While the elderly members of the upper caste family would prefer not to eat the food offered by lower castes, youngsters do not show much hesitation. It is found that interdining, however, is not fully accepted by the upper caste members. In some instances, Lingayats and Gowdas are served separately cooked vegetarian food in a place different from where the function is organised or in an upper caste household. Nagamma, who is from the Waddar community, which claims high social status among Dalit groups, noted that even though Lingayats recognise their higher social status and allow them to enter their temples, they would not prefer to eat with them. She notes:

> We arrange separate food for them (Lingayats, Gowdas and Kurubas). Our caste (Waddar) people will eat from here only. For other big people (Lingayats), if we want to invite them, we have to arrange separate food for them in Kurubara galli (which is outside their street). We prepare *rotti* (from bajra and jowar flour), *pundi palya* (prepared with spinach), potato and drum stick curry for us. But for upper castes, we prepare chapathi, sweet and *khara boondi* (a kind of savoury).

While there are changes in caste practices, the strongly embedded norms of domination and subordination, which were internalised by the Dalits over a period of time, continue to play a role in normalising some of the discriminatory caste practices. Such normalisations are reflected in all spheres of life including worship and public festivals, where caste is being practised along its *graded* hierarchical relationships in several forms. Articulations of such discriminatory practices came mostly from those who could not move out of the traditional caste-based occupations. It is important to mention here how they normalise discrimination in public places like temples. While the village has separate temples for each caste group, important Hindu religious festivals are organised mostly around Veerabhadra and Hanuman temples (the two major places of worship in the village), which are managed by upper castes. There is no apparent restriction for the entry of any Hindu group to the temples. While people belonging to the Waddar community (touchable Dalits) freely enter these temples, it is found that Holeyas and Madigas still hesitate to enter for the fear of resistance from Lingayats and the *dosham* (curse) that they think would be placed on them by the gods if they pollute the temples with their presence. What signifies pollution for them often is the "non-vegetarian food" that they consume.

Though not remarkable, changes have also started appearing in the political spheres of the villages, particularly regarding participation in governance and other political processes. North Karnataka, especially districts like Bidar, Bijapur and Gulbarga which share borders with Maharashtra, is a region that witnessed a strong political mobilisation of Dalits. Dalit movements and activism have been strong in the region. Groups like Karnataka Dalita Sangarsha Samithi, Karnataka Janandolan Sangha, Karnataka Dalita Kriya Vedike and Dalit Sena, are also present in Meenkera, although they are not very active at present. It was found that participation in political and electoral processes helped the Dalit middle class to come to leadership. Similarly, Dalits, especially youngsters from Holeyas and Madigas in Meenkera, are capable of exercising their voting choices in elections. This has also weakened the influence of Lingayats in the larger political landscape of the village, where they acted as a force for ensuring voting blocks for a particular party. Rajesh, who is an educated young man from Madiga community, explains the changes as follows:

> The entire village used to obey the leader, who used to be an upper caste person. People earlier used to cast their votes only to the candidates whom he suggested. Our people (referring to Dalits) used to obey it. They thought that Dhani (leader) suggested this person for our good. But now, our people take their own decisions. Though leaders still try to dictate to our people, they say 'ok' to them but think twice before voting.

While participation in the electoral processes showed such "positive" changes, they did not seem to have been reflected in the local level political processes. The reservation in local self-governing institutions has technically included Dalits and women in the decentralised governance processes. We have found that Dalit members get elected to the reserved seats in Panchayats and Dalit men participate in ward committee meetings and Gram Sabhas. However, as already reported from Karnataka and other parts of India (Inbanathan & Sivanna, 2010), they are often unable to represent or articulate their concerns.

Making sense of *many narratives*

As these narratives show, caste practices have shown resilience in several spheres in Meenkera as well. It has shown changes in the economic organisation of the village as several Dalits could move out from older forms of caste hierarchies. Some of the educated and employed young Dalits even denied the existence of any form of caste-based discrimination against them. Also, caste consciousness is very strong among young and educated Dalits, which led to assertions and questioning of caste practices, if any. While there were changes, caste continuities were also visible. Even though direct economic dependency on Lingayats and Gowdas declined for some Dalits with the moving out to non-agriculture occupations, it was found that such avenues were also controlled by the upper castes, which in turn created new forms of obligations and dependencies. Similarly, some groups from Dalits, who have been traditionally loyal to upper castes like Waddars in Meenkera, continue to be attached to their lands through contractual relationships. The availability of non-agriculture employment and public works helped Dalits to come out of long-term contractual employment relationships with Lingayats and Gowdas; however, the uncertainties of such jobs forced at least some of them to continue to be loyal to these landlords. Some of those who are in public employment and those who moved out to urban places also confronted caste at the workplace and place of residence though they enjoyed respect in their village. While the caste relations got reproduced in the official interactions for public servants, Dalits who migrated to urban places were caught up again in lower level jobs in the informal sector and caste-based cleaning occupations in the cities.

Although caste practices showed a decline in public places, they continue in several forms in private spheres, with restrictions on interdining and visibly different places of habitation and worship. Sometimes, Dalits cannot challenge such practices and they have taken up a strategy to keep themselves away from places and instances which they consider are discriminating. In some instances, Dalits appear to have accepted the notions of *cleanliness* and *bhaya*, which upper castes often use to legitimise their discriminatory behaviours at their residence and temples. Although the traditional powers of Lingayats and Gowdas as rulers of the village declined, they remain controllers of the local self-governing institutions, which also leads to the exclusion of Dalits in equal participation and from the benefits of various welfare activities and development programmes being implemented through local self-governments.

What is emerging strongly from the narratives of Dalits from Meenkera is that there is no single story of change. The experience of caste for Dalits varied notably across persons, individual castes and occupational groups, with further sub-divisions, depending on the situation in which they confronted caste. This has formed their individual understanding of change in caste relations. For older generations, what they witness these days such as entry to upper caste settlements, temples and their houses are *big* changes. Similarly, they attribute negotiating for wages; confrontations and participation in local governing institutions are major changes. Some of the youngsters in the village who are educated and alert to citizenship rights, sometimes from the same caste group or even family, however, consider these as a reflection of caste in everyday life in various forms. For them, any behaviour, which is less than equal, is discriminatory. This group therefore takes a position that caste continues without any significant change in its content. The premise and reference point of change is important here. While the latter premises it in the possibility of equality and equal opportunities, the former sees it from their already accepted inferior positions in the caste hierarchy. The important question related to mobility and change therefore here is how we view it – from a point of time in a stratified social order with a lens of caste hierarchy or from a point of view of a possible more equal social order. If we choose the first one, there are several changes and continuities. The second one, however, points to several changes, which are to be brought about.

Another important dimension of change relates to the difference in caste experience at a personal level in various spheres. For instance, while a *good* job in public or private sector provided Dalits dignity and respect in the village, they experienced discrimination from their upper caste colleagues in other spheres – for instance at the workplace. Similarly, the possibility of educated and assertive young Dalits being explicitly discriminated against in their village is less; they continue to face discrimination from similarly educated upper castes from outside their village in various direct and indirect forms once their caste identity is revealed. It implies that the respect due to education, employment and economic status that Dalits gain become possibly limited only to places where they have a relative advantage over other castes in these attributes. It also indicates that caste can still take several forms depending upon the place one is in within the power relations of caste.

Concluding remarks

It is true that the narratives of Dalits from various socio-economic demographic and aspirational groups in Meenkera also were in tune with the available explanations of caste. There were instances of continuity of hierarchies in socio-economic and interpersonal relationships, complete change of caste practice in public space, incremental changes, changes by assertions and confrontations and changes that could be due to the independence of Dalits. However, the experiences of caste were not common even for an individual across all spheres and situations. Caste, in this particular context, crystallises more at an individual level as its power dynamic has already come out of organised institutional structures to accommodate itself to *spaces* and *situations* where it is more vague and pervasive. This implies that there are no common patterns in the experience of caste even for a village, an individual caste, a family and an individual, which do not allow us to attribute change in caste relations as any convenient category, which may be true only for a given situation. It also contests the understanding of the common pattern and direction of change and mobility for Dalits through various processes such as education, self-employment, political mobilisation, modern employment, etc. as presented by the large-scale empirical research. This chapter shows that favourable conditions of change such as numerical dominance, education and non-agriculture employment did not lead to a common pattern of change for all Dalits in Meenkera; rather, these have benefited only a few individuals from the community. Such changes therefore are real only for those particular persons. Hence, one cannot give common meanings to caste relations, and explanations of change based on available categories should be limited only to its corresponding spheres and situations.

Note

1 This chapter is based on a research project sponsored by the Indian Council of Social Science Research, New Delhi. I thank Shrinidhi Adiga, Lucky Prithviraj and Hussain Sab for their support in conducting the fieldwork.

References

Armstrong, N. (1998). Checks to integration: AKs of Mahepura. In S. R. Charsley & G. K. Karanth (Eds.), *Challenging untouchability: Dalit initiative and experience from Karnataka* (pp. 154–186). New Delhi: Sage Publications.

Assadi, M. (2004). Social movements in Karnataka. In J. Desrochers & P. V. Veliyannor (Eds.), *Poverty, marginalisation and empowerment in Karnataka* (pp. 197–215). Bangalore: KRCR-JPC Publication.

Charsley, S. (1998). Increasing autonomy: The Harijans of Rateyur. In S. R. Charsley & G. K. Karanth (Eds.), *Challenging untouchability: Dalit initiative and experience from Karnataka* (pp. 212–239). New Delhi: Sage Publications.

Deshpande, A. (2011). *The grammar of caste: Economic discrimination in contemporary India*. New Delhi: Oxford University Press.

George, S. (2016). *Discrimination, health access and health seeking behaviour: A study of Dalits and Muslims in selected villages of Karnataka* (Unpublished Research Report). Bangalore: Institute for Social and Economic Change.

Gupta, D. (2000). *Interrogating caste: Understanding hierarchy and difference in Indian society*. New Delhi: Penguin.

Gupta, D. (2005). Whither the Indian village: Culture and agriculture in 'rural' India. *Economic and Political Weekly*, *40*(8), 751–758.

Guru, G. (2016). Shifting categories in the discourse on caste and class. *Economic and Political Weekly*, *L1*(47), 21–25.

Harriss-White, B., & Basile, E. (2014). *Dalits and Adivasis in India's business economy: Three essays and an atlas*. New Delhi: Three Essays.

Inbanathan, A., & Sivanna, N. (2010). *Scheduled castes, legitimacy and local governance: Continuing social exclusion in Panchayats* (Working Paper No. 257). Bangalore: Institute for Social and Economic Change.

Jodhka, S. S. (2002). Caste and untouchability in rural Punjab. *Economic and Political Weekly*, *37*(19), 1813–1823.

Jodhka, S. S. (2010). Dalits in business: Self-employed scheduled castes in North-West India. *Economic and Political Weekly*, *45*(11), 41–48.

Kapur, D., Babu, D. S., & Prasad, C. B. (2014). *Defying the odds: The rise of Dalit entrepreneurs*. New Delhi: Random House.

Karanth, G. K. (1996). Caste in contemporary rural India. In M. N. Srinivas (Ed.), *Caste and its twentieth century Avatar* (pp. 87–109). New Delhi: Penguin.

Karanth, G. K. (2004). Replication or dissent? Culture and institutions among 'untouchable' scheduled castes in Karnataka. *Contributions to Indian Sociology*, *38*(1 and 2), 137–163.

Karanth, G. K., & Ramaswami, V. (2005). *At loggerheads or towards sustainability? Changing rural livelihood systems and natural resource management* (Monograph). Bangalore: Institute for Social and Economic Change.

Madheswaran, S., & Attewell, P. (2007). Caste discrimination in the Indian urban labour market: Evidence from the National Sample Survey. *Economic and Political Weekly*, *42*(41), 4146–4153.

Manor, J. (2007). Change in Karnataka over the last generation: Villages and the wider context. *Economic and Political Weekly*, *42*(8), 653–660.

Manor, J. (2012). Accommodation and conflict. *Seminar*, (633), 14–18.

Martin, N. (2015). Rural elites and the limits of scheduled caste assertiveness in rural Malwa, Punjab. *Economic and Political Weekly*, *50*(52), 37–44.
Mayer, A. (1996). Caste in an Indian village: Change and continuity 1954–1992. In C. J. Fuller (Ed.), *Caste today* (pp. 32–64). New Delhi: Oxford University Press.
McMillan, A. (2005). *Standing at the margins: Representation and electoral reservation in India*. New Delhi: Oxford University Press.
Ram, N. (1995). Social mobility and status-identification among the scheduled castes: A synoptic view. In K. L. Sharma (Ed.), *Social inequality in India: Profiles of caste, class, power and social mobility* (pp. 440–458). Jaipur: Rawat Publications.
Shah, A. M. (2007). Caste in the 21st century: From system to elements. *Economic and Political Weekly*, *42*(44), 109–116.
Srinivas, M. N. (1994). *The dominant caste and other essays*. Delhi: Oxford University Press.
Srinivas, M. N. (2003). An obituary on caste as a system. *Economic and Political Weekly*, *38*(5), 455–459.
Teltumbde, A. (2008). *Khairlanji: A strange and bitter crop*. New Delhi: Navayana.
Thorat, S., Mahamallick, M., & Sadana, N. (2010). Caste system and pattern of discrimination in rural markets. In S. Thorat & K. S. Newman (Eds.), *Blocked by caste: Economic discrimination in modern India* (pp. 148–177). New Delhi: Oxford University Press.

7

CASTE, RELIGION AND RECOGNITION

Trajectories of Pasmanda Muslim movements

Tanweer Fazal

The Pasmanda movement in this chapter broadly refers to various struggles for recognition, and resistance against elite hegemony being waged by the Muslim middle and low castes in different parts of the country. Despite their claims of an all-India profile, these movements are largely confined to certain states of India, and typically correspond with the government's adoption and extension of the policy of social justice to its Muslim populations.

In this chapter, I restrict my analysis of the trajectories adopted by the movement (or movements) in two states of India, namely Bihar and Maharashtra. The policy of reservation for the Backward Classes and instrumentalities of their implementation in these two states are somewhat identical. Both state governments recognise castes (and not community as in certain other states) as units of social and material deprivation and have gone ahead with bifurcating the OBC into sub-categories ostensibly to evenly distribute the benefits of quota amongst the advanced and marginalised social groups. In terms of their origins, the movements in Maharashtra emerged with the formation of the All India Muslim OBC Organisation under the leadership of Shabbir Ansari in 1978, while those in Bihar rose a decade and a half later in the 1990s, spearheaded by the All India United Muslim Morcha (1993) and Pasmanda Muslim Mahaz (1998). In many respects, the movement echoes the politics of Mandal and Backward Classes that seized north India in the 1990s and much of southern India even prior to Independence.

The movement privileges the question of caste differentiation among the Muslims of India, and in a certain way, offers a critical shift from the community-centred theorisations of deprivation and discrimination that post-Independence Muslim politics appeared to be absorbed with. The *ashraf*, *ajlaf* and *arzal* classification, now an indubitable social reality, informs the ideological frame and the theory of emancipation that the movement seeks to endorse. There is ample literature on the prevalence of descent-based segmentation among the followers of Islam in India. Beginning from the late nineteenth century, the earliest exercise of enumeration noticed the existence of caste and a caste-like social hierarchy that also found resonance in the accounts of colonial officials, missionaries and in the imperial gazetteers. The Census of 1901, for instance, in its Report on the Bengal provinces detailed the practices of segregation maintained by the Muslim "functional groups" in terms of rules of commensality, endogamy, burial of the dead, ostracisation of deviant members, authority of caste panchayats, etc. (Census of India, 1901). For E. A. Gait, the superintendent of the Bengal Census, the affirmation of foreign descent and high birth claimed by the Muslim ashrafs mirrored the status of the *dwijas* (twice-borns) in the Hindu hierarchy. However, the colonial production of knowledge of castes and status groups in Islam rarely translated into policies and statecraft. Instead, from the early twentieth century onwards, minority protectionism and safeguards became the mainstay of the colonial state's approach towards Muslim society.

Sociological studies on Muslim social stratification largely followed the ethnographic details and classification offered by the colonial censuses. Thus, the ashraf–ajlaf dichotomy came to be particularly favoured by scholars invested in studying Muslim social stratification through available archival data. In contrast, the field view of Muslim social stratification found the framework over-simplistic as it ignored the smaller units analogous to castes in Hindu society.

These "caste-analogues", according to I. Ahmad, were the real units of social stratification that determined "day to day relationship between different individuals" and thus were the real operational units characterised by endogamy, hereditary membership, specific style of life, traditional occupation and a "distinct ritual status in a hierarchical system" (Ahmad, 1967). The tripartite distribution of Muslim groups was not universally valid across all regions, especially in southern India. Mattison Mines's study of Tamil Nadu argued that the hierarchy among Muslim groups was more related to the level of socio-economic development of the group than their position in the ritual hierarchy (Mines, 1978).

In contrast to the cultural and the ideological perspective that saw caste as a phenomenon limited to Hindu societies, sociologists studying Muslim social stratification adopted the structural approach. It allowed them to look beyond the *dharmashastras* – the divine theory of origin of Varnas in Rig Veda, the ideas of purity and pollution in *Manusmriti* and the theory of karma in the Upanishads – that offered ritual validation to the practice and preponderance of caste in Hindu societies. The structural approach was expediently deployed to resolve the persistent conundrum that emerged from anthropologists' observation of everyday practice of caste among South Asian Muslims on the one hand, and the near absence of its ideological endorsement in Islamic scriptures on the other. Zarina Bhatty applied the basic attributes of Hindu castes to comprehend Muslim social stratification in a village in Uttar Pradesh and found that despite the absence of ideas of purity and pollution in scriptural Islam, "the degree of impurity of the occupation" was a major determinant of caste ranking. Her conclusions rested on structural conditions rather than on ideological explanations: "the Muslim caste structure had its roots in a politico-economic system which was feudal" (Bhatty, 1996).

The tension between the ideological structure of doctrinaire Islam and the social reality of Indian Muslims offers a paradox for social scientists studying the subject. This has led to a multiplicity of explanations. One of these has been to explain the social order as a product of the "dynamic interplay" between ideology and historical forces. The attempt has been to demonstrate how hierarchy not only informed the social organisation of Muslims, but also came to be legitimated through particular interpretations of scriptures and traditions. In this regard, the principle of *Kifa/kafaah* or instructions to establish equivalence of status in marriage is cited as one of the bases for controlling social relations and perpetuate ranking of prestige (Ahmad, 1967). Reference is also made to the system of distribution of positions in the courts of Muslim rulers that preferred foreign descent for high offices. For sociologists deriving the fundamental postulates of caste order from Hindu scriptures, its prevalence in non-Hindu religions is seen as a weak link that developed primarily in a social environment that was predominantly Hindu. Thus, caste among Muslims is argued to have emerged as a result of a prolonged contact with Hinduism and a cultural environment influenced by it. For Dumont (1980), therefore, caste among Muslims is weakened or incomplete but not lacking altogether.

This dissonance between ideology and the social reality of caste also allowed flexibility in rules of observance. While the initial impression betrays a rigid structure akin to the strict rules of adherence in the Hindu caste order, in actual practice, considerable deviations existed. Thus Dumont notes how the transgression of endogamy rules amongst the castes within the ashraf category was socially permissible. Although rare, taking a second wife from one of the non-ashraf castes was accepted in the areas of Bengal Presidency, the 1901 census recorded. Apart from place in the social ranking, what also mattered was the economic status of those entering into marital alliances. An ashraf family, which had fallen on bad times, would be open to considerations of forming an alliance with a wealthy ajlaf family. New-found wealth and prosperity allowed families placed lower down the social hierarchy to lay claims over higher status. Thus akin to Sanskritisation, *ashrafisation* or *Islamisation* has been a process whereby upwardly mobile members of non-ashraf groups claimed high social status. The entire process was gradual and fraught with tensions between the entrenched and the parvenu groups.

The 1901 Census of Bengal explained the process through the following example:

> A well-to-do man of a functional group, say a Jolaha, will discard the word Jolaha, call himself Shekh and assume a more respectable name, and if he be hospitable, he will slowly secure for himself a circle of friends from the poorer classes of the Ashraf community. He will then marry into an Ashraf family, probably of doubtful status, and his son will be recognised as a true Ashraf. The process is easier in the towns than in the villages, where people are more conservative, and in a distant place than near home, where a man's connections are well known. The greatest difficulty that a parvenu has to overcome is to gain recognition for his female belongings from the Ashraf women. The latter are much less easily won over than the men, and will often decline to hold intercourse with women of lower rank, even when very wealthy.
>
> (Census of India, 1901)

Studies of caste and social stratification among Muslims of India have generally been absorbed with the centrality of caste or caste-analogues in comprehending Muslim social structure, the extent to which caste among Muslims and Hindus mirror each other, the question of Islamic egalitarianism as against prevalent social hierarchy, the enforcement of endogamy, ideas of purity and pollution, identifying the process of ashrafisation and provisions of upward social mobility and so on. With its focus on largely the cultural dimension of caste among Muslims, the interface between caste and politics has largely been left unattended. This chapter studies the mobilisation of castes in the domain of movement politics among Muslims of India. In the process, it foregrounds the ideological framework, the theoretical premise, the inherent ambiguities and tension, and the trajectory of the Pasmanda movements.

The chapter intends to provide a structural analysis of the emergence of a caste-centric movement amongst the Muslims of India. Caste-based political mobilisation among Muslims has a history that takes us to the late colonial period when caste/biradari associations were formed to protect the interests of specific groups. Thus organisations such as Jamiat-ul Ansar (of weavers) and Jamiat-ul Quresh (butchers and meat sellers) were formed in the 1920s. Around the same time, Momin Conference, a formation of the weavers of eastern UP and Bihar, was floated as a political front that contested elections against the Muslim League with a certain degree of success. Although some factions of the contemporary movement draw inspiration from earlier attempts to consolidate caste consciousness among the Muslim low castes, the contemporary Pasmanda movement differs from them both in form and content. The major point of departure is in the attempt to build a caste federation and forge a collective consciousness around a presumed common experience of humiliation and deprivation. The commonality of Pasmanda experience, the movement leadership claims, cuts across caste as well as religious divides as is evident in the slogan *dalit pichhda ek samaan, kya hindu, kya musalmaan* (dalits and backwards are one and same, whether Hindu or Muslim).

Ideology and legitimacy

The prevalence of status-based social hierarchy among Muslims akin to the Hindu caste system has been foregrounded by numerous sociological studies, gazetteers and colonial censuses. However, as a significant point of departure from the Hindu caste order, the movement leaders refer to the missing scriptural and ideological validity for descent-based social ranking and prejudicial practices in doctrinaire Islam. This dimension of the Muslim caste order is maintained in their utterances as well as in the documents produced by them. The stress on the absence of religious legitimacy is a critical component in the movements' ideological and political posturing. Not only is their quest for equality (*masawaat ki jung*) presented as in consonance with the precepts of Islam, the dominance of the adversaries – the ashrafs – is illegitimated as antithetical to religion's fundamental injunctions by the movement. The Pasmanda movement has thus been able to develop a scathing critique of the Muslim clergy for its occasional sanction and continued silence over caste practices. The movement is also, according to Shabbir Ansari, the leader of Maharashtra movement, a quest to recover the goals and objectives of true Islam.

By mobilising and invoking caste identities, the leaders and movement intellectuals hope to confront the danger of communalism and the rise of Hindu nationalism that India has witnessed since the decade of the 1990s. In this regard, the prescription that the movement offers aims at fracturing the solidarities of religion by the mobilisation of oppressed castes. The invasion of vertical identity of religion with the perceived horizontality of caste groups similarly placed in the Hindu and Muslim social structure is assumed to be the most effective means to combat both majority and minority communalisms. In a nutshell, it prescribes a competing politics of recognition, caste as against religion. By forging a unity of action and purpose with the social groups similarly placed in the Hindu social order, the PM hopes to build a majority of the oppressed – the *bahujan*.

In the PM discourse, communalism is essentially elite mobilisation for pursuing sectional interests and the appropriation of resources. On this count, majority and minority communalisms are held indistinguishable. On many occasions, Pasmanda ideologues have suggested a conspiratorial mutuality of interest between Hindu nationalists and Muslim community and religious leaders. The violence against religious minorities is produced out of this symphony. In certain anti-minority violence as in the case of Muzaffarnagar, 2013, in which around 45 people lost their lives – a majority of them being Muslim – movement enthusiasts held the Muslim leadership responsible for instigating and even organising it. In such violence, they would argue, it is the subaltern Muslims who face the brunt, while the elite Muslim leadership is able to exploit it to consolidate its political position within and outside the community.

Having positioned itself at variance from the Muslim "elite" politics, the PM's engagement with the dominant nationalist narrative is ingenious. While staying clear of confronting the majoritarian paradigm, the movement ideologues seek adjustment within it. In this regard, the separatist politics of the Muslim League, which according to nationalist historiography, led to the partition of the country in 1947, is denounced as primarily a product of Muslim ashraf. To prove their patriotic claims, the movement leaders have relied on instances where leaders and organisations of Muslim low castes offered resistance to the idea of Pakistan. In this regard, certain subaltern groups such as the Meos of Mewat trace their patriotic history to a medieval past, when Meos and their chieftain apparently braved the invading Mughal armies. The cultural binary of Indic and non-Indic religions that is often deployed by Hindu nationalists against followers of Islam and Christianity is sought to be overcome by referring to the indigenous origins of the Muslim converts as against the foreignness of the ashraf castes.

Strategy and tactics

The centrality assigned to caste identity in the Pasmanda movement takes it away from constructs such as "minority" or minority rights that Muslim politics for a long time pursued. In the words of leaders such as Ali Anwar, "Pasmanda" is the term that is closer to the material world of a large majority of Muslims belonging to the lower social strata. In contrast, the category "minority" obfuscates the reality, presents Muslims as a monolithic community and has allowed influential Muslims to corner all its social benefits. Pasmanda, a Persian word for the deprived, for the movement activists, is a powerful political tool that would eventually demolish all vested interests and empower the powerless.

As the term Pasmanda is deployed in contradistinction to the category "minority", it follows that the movement has been vociferously opposed to policy initiatives targeted towards addressing the deprivations of minorities as a whole. The resurgence of demands for a minority quota following the revelations of the Sachar Committee (Govt. of India 2006) has been vociferously refuted by the leaders of the movement. Such a scheme, it is argued, ignores the inequality and differentiation within, and bears the risk of being appropriated by the elite ashrafs. Questions of religious discrimination and deprivations emerging from majority–minority hierarchy are considered an inaccurate presentation of a far more complex social reality in which status differentiation continues to be the primary basis for the allocation of material and intellectual resources. The movement therefore tends to adopt the "caste is class" kind of argument that most caste-based social and political movements in India subscribe to.

While the movement posits to confront ashraf hegemony in the realm of the social and the cultural, it has, for the most part, restricted itself to seeking recognition from the state as OBC and Scheduled Caste Muslims. The Pasmanda subsumes both Muslims belonging to OBC as well as the Dalit category. Sociologically, this brings both the ajlaf (converts from "clean" Shudra or peasant and artisanal castes) and arzals (converts from "unclean" Dalit communities) together on a common socio-political platform. Since the Presidential Order of 1950 excludes the Muslim and Dalit Christians from the ambit of Scheduled Castes, Muslim groups with similar inherited occupations as those of SCs have found inclusion in the category of OBC. This has led to protests demanding the inclusion of the latter within the Scheduled Castes. Simultaneously, the Pasmanda politics is oriented towards safeguarding the OBC status solely for ajlaf and arzal Muslims. With its focus primarily on social backwardness manifested in the prestige ranking, the movement has typically thwarted the attempts by Muslim groups, purportedly from higher castes, to avail the OBC status.

The trajectories of the movement

The movements in Maharashtra and Bihar, despite a certain commonality in their origins and aims, traverse distinct trajectories. These differences can be noticed in the realms of strategy, politics and ideology. This nuanced approach also helps us in understanding the ideological differences between the Pasmanda and the Dalit Muslim movements in Bihar and other parts of the Hindi belt that have gradually begun treading different paths.

The movement in Bihar

Muslims in Bihar, like elsewhere, are stratified into ashraf, ajlaf and arzal categories. Since arzals are not recognised as SC, they have been included as OBC and Most Backward Caste (MBC) in the state. Prior to the implementation of the Mandal Commission report (1989), the Mungerilal Commission instituted by the state government recognised nine Muslim groups as OBC and 27 as MBC. A large majority of Muslims of Bihar declared their OBC status – 65.6 per cent during the 2009–2010 NSS count. This has further increased to nearly 76 per cent in 2011–2012, suggesting a high level of status consciousness among them. Of late, there have also been attempts by certain Muslim high castes to get themselves included within the OBC category.

It is argued that the emergence of a Pasmanda consciousness – a consciousness based on commonality of the experience of marginalisation and deprivations amongst different castes and social groups – has remained a far cry. Various Muslim caste and occupational groups, be it the *momins*, the *quraishis*, the *rayeen*, the *hawaireen*, the *abbasis*, etc. maintain their distinct social organisations and occasionally pursue sectional interests. In Bihar, for instance, since the 1980s, the momins (weavers) and the rayeens (vegetable-sellers) have rarely seen eye to eye. The rivalry dates back to the 1980s when the momins, by virtue of their political clout, got themselves included in the EBC (Extreme Backward Classes) list of the state, while the latter's demands were left unattended. The existence of the two different Muslim OBC platforms in Bihar, the AIBMM (rechristened as AIUMM) and PMM, is a manifestation of the existing split in the movement. Despite aggressive anti-ashraf rhetoric, the two organisations differ in terms of their reading of history, in their choice of movement icons and more so, in comprehending and pursuing the agenda of social justice.

At the conceptual level, the attempt to absorb a Dalit identity within the all-inclusive signifier of marginality and caste discrimination – the Pasmanda – has come to be seriously interrogated by activists and movement ideologues espousing the cause of Dalit Muslims. The movement has increasingly come to adopt the hyphenated identity of Dalit–Pasmanda that suggests the immutability of arzal identity within the ajlaf. In Bihar and to a certain extent in Uttar Pradesh where the movement drew considerable intellectual attention, this divide between the Dalit and the Pasmanda is wide open. The AIBMM, the rival organisation, positions itself as the representative body of Dalit Muslims in India. In the year 2008, Dalit Muslim leaders broke away from PMM, alleging appropriation of the movement by the dominant middle castes. The immediate provocation for the dispute was also the question of communalism and how the Pasmanda movement in Bihar would approach this. As Ali Anwar, the founder of the PMM, accepted his nomination as an MP of ruling JDU in Bihar, then a constituent of BJP-led NDA, PMM members saw it as a moment of compromise of the movement with the communal BJP. A similar fate was awaiting the other organisation AIBMM too as Ejaz Ali, its leader, accepted JDU nomination.

The emerging Dalit–backward schism in Pasmanda politics follows the pattern observed in the wider social justice movement in India. The conflict between the Marathas and the Mahars in Maharashtra, the Dravidas and the Adi-Dravidas in Tamil Nadu or that between the OBCs and the various Dalit castes such as the Chamars in north India has been well documented. But much of this conflict has been explained in terms of emerging agrarian relations and the rise of new agrarian classes. It has been argued that the zamindari and tenancy reforms in India produced piece-meal social mobility, at the most, at the interstitial strata of agrarian caste–class structure. The abolition of large estates and the imposition of land-ceilings allowed certain agrarian castes to join the ranks of the kulaks. In the absence of radical land redistribution, those at the bottom of the class hierarchy, namely the low caste field servants, continued to bear the brunt of the exploitative system. The increasing tension between the Dalits and the OBCs is partly attributed to this new relation of exploitation in the agrarian structure.

The framework of changes in agrarian relations to explain the emerging Dalit–Pasmanda divide in the social justice movement is difficult to explain. This is primarily because of the nature of class formation among Muslims. While the Muslim occupational structure has its own regional specificity, at the all-India level, a pattern distinct from the general population profile emerges. Muslims in the rural areas are mostly self-employed; however, this doesn't suggest greater incidence of self-cultivation and landholding, but rather pauperisation. Compared to all others, Muslim self-employment in agriculture is the weakest, 21.4 per cent as compared to 33.2 per cent being the general pattern. Most relevant to our analysis is the pattern of agricultural self-employment within the OBC category. While nearly one-third of non-Muslim OBCs are self-cultivators or owner-cultivators, amongst the Muslim OBC, this comes down to less than one-fifth of the rural population. In fact, a comparatively larger proportion of Muslim OBCs are self-employed in non-agriculture activities – a feature that they share with the Muslim upper castes (Fazal and Kumar 2018). These occupations presumably include petty economic activities such as beedi making, basket making, being washermen and barbers that the rural economy offers.

In contrast to the general pattern, the bulk of the Muslim population in Bihar (87 per cent) resides in the villages as do most other residents of the state. However, a majority of the Muslims are either agricultural labour (39 per cent) or are self-employed in non-agricultural activities (21.5 per cent). In fact, after SC/STs (46.1 per cent), the proportion of agricultural workers among Muslims is the highest. Agrarian relations in Bihar do suggest a pattern of consonance between caste and class locations as upper castes tend to be found more in the category of landlords and owner-cultivators while vast numbers of field-workers come from the lowest castes. This pattern of relations is not exactly repeated in the case of Muslims. Nevertheless, Muslim upper and lowest castes do have differential holdings. A much higher proportion of Muslim OBCs, which includes the Dalits and the MBCs, constitute the agricultural labour (45.4 per cent) while among the UCs, the fraction is much smaller (25.5 per cent). Muslims as a whole have the least self-employability in agriculture, just a shade higher than the Scheduled Castes; nevertheless, within the community; it is the upper castes who return higher figures. In terms of the size of land holdings again, a larger proportion of Muslim OBCs tend to have much smaller holdings. Nearly 50 per cent of them are just marginal farmers while nearly 28 per cent of Muslim UCs can be counted amongst marginal farmers. Compared to the OBCs, a much higher proportion of Muslim UCs is small (64 per cent as against 45 per cent) and middle farmers (7 per cent as against 2 per cent). The presence of Muslims, of either category, among rich farmers is negligible.

Overall, when compared with the other socio-religious communities, Muslims of Bihar remain the most impoverished. The incidence of poverty is disproportionately high among Bihar Muslims. As per the Sachar Committee findings, a very large section of Bihar Muslims are impoverished – 45 per cent in urban areas and 38 per cent in rural areas. A survey of Bihar Muslims found only 35.9 per cent of Muslim households in rural Bihar possessing any cultivable land, which was much below the state's average of 58 per cent landowning households in rural Bihar. Amongst landowning Muslims, for nearly one-fifth of the households, the size of the holding was so uneconomical that they preferred leasing it out to other cultivators while remaining dependent on either wage employment or self-employment (Bihar State Minorities Commission, 2006:19). While the average landholding in Bihar is supposedly 2.32 acres, the Minorities Commission survey found the average size of holdings to be much lower for Muslims at 1.91 acres. Only 8.2 per cent of the Muslim households had landholdings over 2.0 acres (ibid.).

The weak landownership pattern among Muslims in general, and Muslim OBCs in particular, suggests that the fruits of zamindari reforms have largely missed the Muslims. The failure of the state in Bihar in the abolition of landlordism and in ensuring land redistribution is well known. However, the state was able to abolish certain large estates, some of which were held by the Muslim aristocracy. This followed the emergence of a new proprietary class of owner-cultivators from among the large tenants. Certain peasant castes such as the Yadavas, the Kurmis, etc. were thus able to join the ranks of the kulak in Bihar, but the same failed the Muslim peasantry.[1] The argument being made here is that post-zamindari land reforms have not altered the agrarian class/caste relations in the Muslim social structure to a degree that it would contribute to the growing conflict among Muslim OBCs and Dalit Muslims. Further, in a purely materialist framework, the grounds of tension between Muslim upper castes and the lower castes seem to have receded with the dispossessions that the former have gone through.

The movement in Maharashtra

In Maharashtra, the All India Muslim OBC Organisation was founded in as early as 1978, primarily focusing on the demand for the inclusion of Muslim OBCs in the state and the central lists. With the Mandal Commission recognising the occupational groups among Muslims, the movement leaders saw it as a window of opportunity to get such groups on a single platform to voice their grievances. Led by Shabbir Ansari, a leader of socialist leanings, the organisation remained primarily an advocacy platform that through protests and petitioning claims to have significant achievements to its credit. When Ansari began organising the backward social groups, he was criticised and admonished by Muslim organisations and their leaders for provoking caste sentiments amongst Muslims. The disapproval came not only from the ashraf elite (nawab and zamindar classes as Ansari prefers to refer to them), but to his surprise, even certain groups who could be counted amongst the ajlaf: "I was physically attacked by a butcher when I sought to convince them how *qureshi* were socially and educationally backward and needed to be included among the OBC. To the *qureshi* self-belief, they had equal claims to being an upper caste." Ansari soon realised that the caste framework had only a limited appeal and failed to address the entire problem of the Muslim social groups.

The All India Muslim OBC organisation, unlike its counterparts in Bihar and Uttar Pradesh, desists from making appeals to caste sentiments as the singular rallying point. In fact, the organisation works along with other Muslim groups including the Jamaat-e-Islami that were otherwise opposed to him. The question of social equity, in Ansari and his organisation's approach, is not separate from the quest for justice and equality within Islam. The Muslim groups that have been listed as OBC in the central and various lists are primarily occupations (*pesha*) that came from artisanal and labouring classes. In the feudal social structure that valued *nawabiyat*, such professions and their practitioners were derided and pushed down the social ladder. They were socially and educationally backward and on account of their backwardness came to be included as OBCs, the organisation leaders explain.

Thus, the theory of social justice that the OBC movement in Maharashtra has developed leaves little room for anti-ashraf rhetoric. Although the movement polemic holds the aristocracy responsible for imposing an unjust and exploitative social order; "*zulm hua hai*, these social groups were denied *izzat* (honour)", according to Ansari, all that happened in the past, when the Muslim elite dominated economy and politics, contemporary social reality is far different. Various studies and reports have proved that *nawabiyat* and zamindari are a thing of the past. Pride in high birth has followed this material decline. However, the movement holds that the backward social groups among Muslims continued to face immense inequities, and needed urgent attention: "We should take up their cause first but cannot leave others behind. Justice loving people should take up everybody's cause (Interview, February 8, 2015)."

Accordingly, the Muslim OBC organisation of Maharashtra has, among its activists and leaders, members of different social categories, including from the ashrafs. Even the Jamaat-e-Islami which had vehemently denounced Ansari, accusing him of introducing casteism and seeds of division within the community, has now started working in tandem with the OBC organisation. Leaders of the Welfare Party floated by Jamaat-e-Islami have lent their support; they have come to realise that development of the *qaum* is impossible without justice to these communities and their needs. Despite more or less an identical history of origin and similarity in terms of the social base, the Maharashtra movement has cast its politics fundamentally in variance from the Dalit and the Pasmanda movement in the Gangetic belt. It spells its discomfort with category such as Pasmanda or Dalit Muslim, both of which, according to them, privilege caste and promote a politics that sharpens such sentiment. This in one sense is another form of politics of identity fundamentally centred around the mobilisation of emotions. Such a politics of recognition ends up ignoring the redistributive agenda that needs to be prioritised, Shabbir Ansari believes:

> Like our elite leaders who were mostly interested in emotional issues such as Urdu, Babri, Shah Bano, the Pasmanda leaders are also raising only emotional issues, identity specific. They are not inclined towards working on social issues, on *bunyaadi* problems. Their interest is *siyaasat* (Interview February 8, 2015).

A conception of justice drawn from Islamic theology informs the politics of the Maharashtra OBC movement to a large extent. Islam, as movement ideologues argue, beckons and commands its followers to adopt the principle of equality and denies legitimacy to tribal, racial or national supremacist ideologies. In their public utterances, the organisers of the movement refer to and draw strength from what is usually referred to as the last *khutba* of the Prophet. The sermon delivered at Mecca called upon Muslims not to "inflict or suffer any inequity". It categorically rejected social differentiations and racial practices: "there is no superiority of Arab over non-Arab, neither the white over the black". This allows the movement to draw immense legitimacy for its pursuit of equity within and between the communities. The movement in Maharashtra has thus been able to enlist the support of the clergy as much as other sections of the Muslim elite.

Though the movement's pursuit of justice is legitimated as an Islamic calling, the movement works on the premise that the roots of the problem of Indian Muslims lie in economic, social and political and not the religious issues. With its focus on the material disparity of the Muslim low castes, the movement claims to have been successful in getting nearly 72 Muslim caste groups recognised as backward in Maharashtra and in the Mandal list. The organization found mentioned in the Mandal commission Report for highlighting the plight of the Muslim backward groups as identical with that of their Hindu counterparts (Govt. of India 1980: 48). The success of the movement can be judged by the fact that the Mandal Commission that submitted its report in 1980 took notice of the movement and commended it for bringing to its notice the plight of Muslim occupational groups.

Apart from the question of representation, they have fought for land rights, educational empowerment, for skill development, for tax waivers to artisans and entrepreneurs, for facilitating bank loans and so on.

In Maharashtra, in contrast to Bihar, Muslims are primarily an urban community with nearly 70 per cent of them populating the cities. Nearly one-third of the Muslim population claims OBC status. While the movement estimates the Muslim OBC population to be around 85 per cent, the fact that a large part of the population refrains from being counted as backward leaves ground for various plausible explanations. In one way, it is a commentary on the extent to which the Maharashtra OBC movement has been able to spread its influence. Second, it could also suggest a high degree of ashrafisation among the Muslim low castes of Maharashtra.

In Maharashtra, empirical studies on the socio-economic conditions of Muslim social groups certainly suggest the absence of any apparent caste–class overlap. The landownership patterns of both Muslim OBCs and others do not show any significant variation. In fact, according to a report by the Maharashtra government, nearly 90 per cent of the Muslims are bereft of any land whatsoever. Neither is there any big difference in the incidence of poverty among the two social strata. The same is repeated if we examine the occupational structure too where both OBCs and non-OBCs are similarly placed in regular salaried works (around 28 per cent of the working population) while the OBCs have a slight edge in self-employment (25 per cent as against 19.5 per cent). However, the non-OBCs tend to be more in unsecured casual work (26 per cent as against 28.2 per cent) and have more numbers of unemployed (15 per cent as against 12.3 per cent) (Govt of Maharashtra, 2013, p. 175).

Even though caste and class do not correspond very neatly among the Muslims of Maharashtra, social divisions and cleavages have existed for long. Broadly, the ashraf–ajlaf dichotomy remains relevant, with the aristocracy claiming higher status and the craftsmen, peasants, artisans and workers occupying lower rank. Rapid changes in social and economic order have also led to substantial shift in the hierarchical arrangement. In Maharashtra, sociologist A. R. Momin drew attention towards the processes of ashrafisation and de-ashrafisation. Certain middle castes, such as Julahas of Bhiwandi, after having acquired prosperity and attained modern education, sought status enhancement in the caste hierarchy. However, we also have evidence of the process of de-ashrafisation when status conscious Konkani Muslims, having lost their material wealth, took to lowly professions and were keen on establishing matrimonial ties with the Ansaris of the region. In other places, Momin could observe converts from Patel castes, theoretically ajlaf, claiming superiority over the *syeds* who generally occupy the highest position in the ashraf-ajlaf schema (Momin 1977).

Antinomies of recognition

The movements for the amelioration of Muslim low castes have had a long history, some of which could be traced to the colonial period. At a broader plane, the movement *weltanschauung* and pronounced goals of the Dalit, Pasmanda or OBC movements appear identical: that is amelioration of the Muslim low caste. However, given the differences over the ideological framework, political trajectory traversed, conception of justice and movement strategy, it is reasonable to argue in terms of a plurality of lower caste movements. In this light, it is argued that the divergences between the movements in Bihar and that in Maharashtra are at once a reflection of the distinct social and political milieu from which they emerged and continue to exist. The structural approach adopted here aims not to diminish individual agency and innovativeness, but offers possibilities to realise the various ways in which the two interact. Movement leaders and ideologues, aware of the structural constraints and challenges, are able to reinvent themselves within the parameters of the broader objectives.

In Maharashtra, the Shabbir Ansari led movement, a pioneer in mobilising Muslim low castes, adopts a non-conflictual approach towards the Muslim elite. With centrality on the question of justice, for which it finds support from religion, the movement has developed a cross-class framework. The reference to religion, however, is philosophical-ideological while the goals are fundamentally material. In Maharashtra, Muslims are primarily an urban community, though a small affluent class does exist, but a large majority of them remain impoverished. Further, a comparatively small proportion of the Muslim population claims OBC status. With their presence in the villages very limited, and no clear agrarian pattern in sight, for the movement to survive on caste antagonism appeared bleak. The Muslim OBC organisation therefore has set the goals of the movement outside the community.

In contrast, the Pasmanda movement in Bihar banks on the centrality of caste, based on which it seeks reparation both within and outside the community. Caste symbolism, anti-ashraf rhetoric and narratives of caste practices are repeatedly deployed to build a pan-Pasmanda solidarity. The movement leaders emphasise on continued disparity within the community such as practices of maintaining separate graveyards, over-representation of the Muslim upper castes in the religious bodies, endogamous practices amongst the ashraf and so on. The PM's quest for intra-community equality has received wide attention and appreciation. But with its goals and strategies lying largely within the domain of politics of recognition, the movement, except for occasional posturing, continues to shy away from redistributive goals.

Further, with its emphasis on securing reservation as the key strategy for meeting the equity needs of the Pasmanda, the movement remains focused on its urban constituents, especially attending to the needs of the emerging Pasmanda middle class. This is also true of its approach towards politics where its electoral strategy has been more or less oriented towards securing a larger share in ticket distribution compared with the Muslim ashraf. The middle class formation amongst the Muslim OBCs of Bihar remains weak and small in size, apparently lacking the capacity to carry a mass movement along.

The focus on representation alone has left a vast section of the Muslim OBCs, residing primarily in rural areas (nearly 90 per cent) largely unattended. As a result of this, Muslim Dalit and Pasmanda Muslims, who are field workers or are marginal farmers, have little reason to find hope within the movement. The tension on Dalit–Pasmanda lines that has emerged within the movement is partly a reflection of the movement's limited social programme. By keeping its goals of justice largely within the community and the social groups constituting it, the Pasmanda movement made itself vulnerable to fragmentations on lines of caste and identity.

The movement in Maharashtra, after having succeeded in getting OBC groups listed within the state and the central lists, has expanded its scope with a vibrant social and economic programme. Aware of the limitations of the recognition framework, it has begun addressing the question of land dispossession, educational deprivations, poverty and unemployment. By setting the goals of justice outside the frame of community, it has been able to enlist the support of Muslims of different social classes. Nonetheless, this also leaves important questions regarding share of positions and resources within the community unattended. Therefore, the casualty in this approach is the task of politicisation of the subaltern castes.

Note

1 Noticeably, among the landowning non-Muslim OBC households, a significant 12 per cent has joined the ranks of the Kulaks with more than 2 ha size of farm land that they own, a category in which Muslim OBCs (or even upper castes) are barely represented.

References

Ahmad, I. (1967, May 13). The Ashraf and Ajlaf categories in Indo-Muslim society. *Economic and Political Weekly*, 887–891.

Bhatty, Z. (1996). Social stratification among Muslims in India. In M. N. Srinivas (Ed.), *Caste: Its twentieth century Avatar* (pp. 244–262). New Delhi: Viking.

Bihar State Minorities Commission. (2006). *Socio-Economic and Educational Status of Muslims in Bihar*. Patna: ADRI.

Census of India. (1901). *Report on the lower provinces of Bengal and its feudatories* (Vol. 4, pp. 438–451). Calcutta: Bengal Secretariat Press.

Dumont, L. (1980). *Homo hierarchicus: The caste system and its implications*. Chicago: University of Chicago Press.

Fazal, T and Rajeev, K (2018). Muslims in India: A Study of Socio-Economic and Educational Levels in Four Focus States, in Zoya Hasan edited, India: Social Development Report, Delhi: OUP, pp. 179–195

Govt. of India (1980), Report of the Backward Classes Commission (Mandal Commission), New Delhi

Govt. of India (2006), Social, Economic and Educational Status of the Muslim Community of India, A Report, PMHLC (Sachar Committee), New Delhi

Govt. of Maharashtra. (2013). *Socio-economic and educational backwardness of Muslims in Maharashtra* (Table 9.6, p. 175). Mumbai: Author.

Mines, M. (1978). Social stratification among Muslim Tamils in Tamil Nadu, South India. In I. Ahmad (Ed.), *Caste and social stratification among Muslims in India* (pp. 159–169). New Delhi: Manohar.

Part III

MARGINALITIES AND DEVELOPMENT

8

YOUTH, ENDOGENOUS DISCRIMINATION AND THE DEVELOPMENT CONUNDRUM

S. Madheswaran, Smrutirekha Singhari and B. P. Vani

Introduction

The persistence of discrimination in the labour market – whether advanced or developing – contributes significantly to social tension and uncertain economic growth. In the case of a developing society, the stake is even higher given that the society is in a transition phase to higher growth and any persistent unstable factor may reverse the process of growth and inject elements of volatility. The Indian economy, unlike many other developing economies, has experienced unprecedented growth over the years, yet there has been constant debate about whether the benefits of such growth vistas have been equally distributed across social strata and demography. Very importantly, the focus of the recent research in academic and policy circles has been whether persistent high growth has ameliorated the problem of discrimination or it has risen along the growth path. It is to be noted that discrimination can manifest itself in all aspects of life such as the type and location of housing available to certain groups, their access to quality education and health care or their treatment in the labour market.

In this context, the present study tries to measure the extent of wage discrimination against Scheduled Castes (SC) in the labour market in India. The quantile regression decomposition method is used to measure the extent of caste discrimination across wage quantiles. This technique helps to find out whether there exists a "glass ceiling effect"[1] or "sticky floor effect"[2] in the labour market. We also seek to address how existing reservation policies for deprived groups have increased (decreased) the effect of caste discrimination, over time. Has the effect of caste discrimination been felt equally across demographic distribution? Has the reservation policy of the Government of India created any significant difference to reduce the effect of caste discrimination between the younger generation and the older generation?

This chapter is structured as follows: the second section discusses the theoretical insights and economic implications of discrimination. The third section reviews the available empirical literature on caste discrimination in the labour market in India. The sources of data and econometric methodology are discussed in the fourth and fifth sections. The empirical results are discussed in the sixth section. The seventh section concludes the chapter and provides policy implications.

Theoretical insights and economic implications of discrimination

Becker's (1971) theory of discrimination, with testable behavioural implications based on a competitive labour market, is often referred to as the "neo-classical theory of discrimination". In this theory, the motivation for discrimination is based on a non-pecuniary variable, generally designated as "taste for discrimination" against a group. For example, employers with a taste for discrimination against Blacks feel that the real burden is more than the money wage burden; the dissatisfaction they feel at the presence of Blacks in their firm is an additional burden. In the neo-classical theory, not only employers, but also employees and consumers can discriminate against a particular group. In India, higher-caste employers discriminate against Harijan ("untouchable") workers.

Defending Becker's theory, Arrow (1972, 1973) defined discrimination in terms of the employer's perception of reality. For Arrow, employers discriminate not because of their "taste to discriminate" but because of uncertainty characteristics of certain group of workers. Phelps (1972) presents a similar discussion, though he prefers to call this the "statistical theory of discrimination", which focuses on the idea that the likelihood that information available to employers about the skill endowments of individual job applicants is relatively sketchy, while information about the average endowments of social groups is (or will be believed to be) relatively complete. The victims of discrimination are generally from low income groups. The average member of a low income group may not be as well endowed with productive characteristics as the average member of the high income group. Thus, in the absence of complete information on individual job applicants, unprejudiced employers may give preference in hiring, or pay higher wages, to members of privileged groups.

There are a few other causes for discriminatory behaviour, like social customs (Akerlof, 1976, 1980) and monopsony in the labour market (Madden, 1975). Akerlof's theory, which deals specifically with caste-based discrimination, assumes that utility depends not only on consumption but also on an individual's prestige and reputation in society, and incorporates the social structure into his model to explain the economic phenomenon of income distribution and resource allocation. Hence, a socially conscious individual would discriminate against the group that prevalent social customs traditionally discriminate against. Birdsall and Sabot (1991) note that an employer's behaviour, if based on social customs, is indistinguishable from that based on taste or prejudice. Invariably, these economic theories of labour market discrimination are based on a micro-economic foundation, and centre on explaining the causes of discriminatory behaviour.

Tzannatos (1987) makes the first attempt to estimate discrimination and its effects on income using a general equilibrium model. The main aim of his study is to find out what would happen to wages and profits if wage differentials by sex were deliberately changed. A few economists have tried to grapple with the issue of economic exclusion and discrimination based on caste and untouchability in India. In the social science literature, the concept of social exclusion is defined as "the process through which individuals belonging to some groups are wholly or partially excluded from full participation in the society in which they live" (Hann, 1997, p. 1). Sen (2000) draws a distinction between "unfavourable exclusion" or situations wherein some people are kept out (at least left out), and "unfavourable inclusion" or situations wherein some are included (even forcibly) under deeply unfavourable terms. Unfavourable inclusion – particularly if the arrangements are unacceptable, or treatment is unequal – may carry the same adverse effects as unfavourable exclusion does. The notion of unfavourable inclusion appears to be quite close to the concept of "market discrimination", which is related to race and gender, in the mainstream economic literature (Becker, 1971).

Economic implications of discrimination

Why are governments in developed and developing countries concerned about economic discrimination? Is discrimination only an equity issue? Does it also involve economic costs to society? Are the costs it imposes on society more social and political than economic? The insights of mainstream economic theory indicate that economic discrimination – and, particularly, market discrimination – hampers economic growth, leads to unequal income distribution and deprivation for discriminated groups, and creates potential for inter-group conflict. Thus, concern about exclusion and discrimination arises not only because it deepens economic inequality and deprives excluded and discriminated groups, but also because it adversely affects economic development. Discrimination also affects productivity as discriminated groups reduce the magnitude of their investment in human capital and, therefore, the return on investment drops (Birdsall & Sabot, 1991).

The consequences of discrimination on income distribution are obvious as far as market exclusion involves denying access to employment and payment of fair and equal wages. In the labour market, groups discriminated against lose income because their wages/salaries are lower than for other groups with similar productivity. Exclusion and discrimination in access to other markets for income-earning assets – such as agricultural land, capital, and other factors – also reduce the income-generating capacity of groups discriminated against. Further, pre-market discrimination in access to education and skill development prevents discriminated groups from raising their level of human development and reduces their chances of employment.

Groups are discriminated against in exchange not only in the labour market but also in other markets – like land, capital, products, and social services such as education and housing – and non-market channels. Such discrimination is a failure of the market, and has consequences. A policy implication of neo-classical theory is that intervention in some form is necessary to correct the consequences of such failure. It is this insight that has induced and justified the adoption of various types of affirmative and positive action policies in favour of discriminated groups in the US and other western countries.

In India, the affirmative action, or anti-discrimination, policy was advocated mostly on the consideration of the violation of citizen/human rights, particularly of low-caste untouchables. Ambedkar, who pioneered the affirmative action policy, based most of his arguments on the human rights perspective, and drew largely from theoretical developments in political science. Ambedkar developed a general theory of caste as far back as in 1916, but its economic underpinnings were elaborated in detail much later in the 1940s and 1950s (and appeared in print only in the 1990s). The efforts to bring about the policy began in 1919, were formalised in 1931 under the Poona Pact, and finally acquired definite legal shape in 1950. Therefore, the discussion on and debate over the affirmative action policy in India has, by and large, been devoid of economic logic or justification. This is in contrast to the large body of literature on the analysis of economic discrimination of race, ethnicity, and gender in the western world (Darity, 1997). Since the main justification for the affirmative action policy in the west emanates from mainstream theoretical economics, the discussion here is limited to this branch of writing. Generally, this theoretical strand agrees on three unique underlying principles and customary rules that govern and structure the production, organisation, and distribution under the traditional caste system:

a) fixed occupation (property rights) for each member of the caste by birth, and hereditary continuation of such occupation
b) unequal distribution of economic and social rights related to occupation, property, employment, wages, education, etc., among different caste groups
c) the provision of a strong system of penalties to ensure enforcement of the system

In terms of consequences, the economic interpretation of the studies by Akerlof (1976), Lal (1988) and Scoville (1991) imply negative outcomes of caste-based market discrimination for economic growth and income distribution. The Akerlof–Scoville–Lal model would thus argue that, given the segmented and imperfect character of the labour market, the economic efficiency of the caste system would be lower than that posited in the model of a perfectly competitive market, and a second-best alternative to the Pareto optimum. Ambedkar (1936, 1987a, 1987b) and Akerlof (1980) pointed out that by restricting the mobility of labour across caste occupations and thereby not permitting readjustment of employment, caste becomes a direct cause of much of "voluntary unemployment" among high-caste persons and "involuntary unemployment" among low-caste persons.

Empirical literature

Azam (2012) and Kijima (2006), using the consumption expenditure data from the rural sector, explain the dynamics of caste discrimination via the distributional effect of consumption expenditure among all social groups. They demonstrate that consumption expenditure is not equally distributed across social groups and a high level of social disparity exists among them. Azam interprets his findings by suggesting that the growing consumption disparities among all social groups may be a potential source of caste discrimination. However, it is important to note that the distribution effects between social groups are not similar to the discrimination effect. In an economic sense, the distributional effect only reflects the lack of access to resources (education, health services, employment, etc.) between two class/groups. Therefore, it is important to understand which resources are helpful in linking two economically deprived or socially separated societies.

Arabsheibani, Gupta, and Mishra (2012) have measured the discrimination effects across social strata, human capital distribution and demographic characteristics. The study finds that despite high growth, discrimination has been on the rise. Specifically, discrimination among the young population across social strata and educational attainment levels is constantly rising, making the economic growth transition path more vulnerable and uncertain than it is commonly assumed to be. They demonstrate that the level of discrimination is not unique across population cohorts; especially, young cohorts are consistently more discriminated against compared to older cohorts and that the discrimination effect is rising over time. The limitation of this study is that it compares two study periods such as 1993–1994 and 2010–2011, which are not comparable in the sense that NSSO's 1993–1994 survey data does not have separate information on OBCs, but the 2010–2011 data has this information. Besides, data collected in 2010–2011 is not in use; instead, 2011–2012 data should be used. Further, the authors discuss the reservation policy without having a separate analysis for public and private sectors in India.

The present study is unique in the sense that it tries to observe the historical trends by examining the issue of caste discrimination by age cohorts over time using different rounds of NSSO's survey data spanning the period 1983 to 2011–2012. We have employed different approaches of the decomposition method to measure the extent of caste discrimination in the labour market. The sophisticated quantile regression decomposition method proposed by Machado and Mata (2005) and Melly (2006) is used to measure the extent of discrimination against SCs across the wage quantiles. The results are compared against the mean-based decomposition method. This measurement of the wage gap across the quantiles of the wage distribution helps to test whether the phenomenon of "glass ceiling effect" or "sticky floor effect: exists in the labour market. We have measured the extent of caste discrimination in public and private sectors separately in order to study the effectiveness of the reservation policy which confines the miniscule arena of public sector and excludes the vast private sector.

Sources of data

The present study uses unit-level data collected by the NSSO, India. The employment and unemployment surveys are conducted during 1983 (January 1983 to December 1983), 1993–1994 (July 1993 to June 1994), 2004–2005 (July 2004 to June 2005), and 2011–2012 (July 2011 to June 2012). These quinquennial rounds are referred to, respectively, as the 38th, 50th round, 61st round, and 68th round. For the sake of convenience, hereafter, we refer to the study periods as 1983, 1994, 2005 and 2012. The survey provides data relating to human capital, demographics, and the job characteristics of workers. Human capital characteristics include age and education; demographic characteristics include gender, social groups, religion, marital status, location (rural/urban), and region; and data relating to job characteristics include industry, occupation, sector, and nature of employment. For more information on the survey and sample design, see NSSO (2014).

The nominal daily wages are deflated by 2001 prices using the official state-level monthly consumer price indices of agricultural labour (base year 1960) for rural wages and consumer price indices of industrial workers (base year 1982) for urban wages (Labour Bureau, various years). The Consumer Price Index data is available for the states like Andhra Pradesh, Assam, Bihar, Gujarat, Haryana, Himachal Pradesh, Jammu and Kashmir, Karnataka, Kerala, Madhya Pradesh, Maharashtra, Orissa, Punjab, Rajasthan, Tamil Nadu, Tripura, Uttar Pradesh and West Bengal.

Econometric methodology

In prior research, one finds three different empirical approaches for studying caste discrimination. The first predicts earnings from the characteristics of all workers while including caste as a predictor (a single-equation technique). Unfortunately, this approach yields a biased result, because it assumes that the wage structure is the same for both FCs and SCs. We have employed the following decomposition methods to estimate the extent of discrimination against SC workers in the labour market in India.

Blinder (1973)–Oaxaca (1973) decomposition method

The decomposition method enables one to separate the wage differential between SCs and FCs into differences that can be explained by differences in characteristics and those that cannot be explained by differences in characteristics. This decomposition can be further applied within the framework of the semi-logarithmic earnings equations (Mincer, 1974) and estimated via OLS such that

$$\ln \bar{Y}_{fc} = \sum \hat{\beta}_{fc} \bar{X}_{fc} + \varepsilon_{fc} \text{ (FC wage Equation)} \tag{1}$$

$$\ln \bar{Y}_{sc} = \sum \hat{\beta}_{sc} \bar{X}_{sc} + \varepsilon_{sc} \text{(SC wage Equation)} \tag{2}$$

where $\ln \bar{Y}$ denotes the geometric mean of earnings, $\bar{X}$ the vector of mean values of the regressors, $\hat{\beta}$ the vector of coefficients and ε is the error term. Within this framework, the gross differential in logarithmic term is given by

$$\ln(G+1) = \ln(\bar{Y}_{fc} / \bar{Y}_{sc}) = \ln \bar{Y}_{fc} - \ln \bar{Y}_{sc}$$

$$= \sum \hat{\beta}_{fc} \bar{X}_{fc} - \sum \hat{\beta}_{sc} \bar{X}_{sc} \tag{3}$$

The Oaxaca decomposition simply shows that Equation (3) can be expanded. In other words, the difference of the coefficients of the two earnings functions is taken as *a priori* evidence of discrimination. If, for the given endowment, SC individuals are paid according to the FC wage structure in the absence of discrimination, then the hypothetical SC earnings function would be given as

$$\ln \bar{Y}_{sc} = \sum \hat{\beta}_{fc} \bar{X}_{sc} \tag{4}$$

Substituting Equation (4) in Equation (3), we get

$$\ln \bar{Y}_{fc} - \ln \bar{Y}_{sc} = \sum \hat{\beta}_{fc} (\bar{X}_{fc} - \bar{X}_{sc}) + \sum \bar{X}_{sc} (\hat{\beta}_{fc} - \hat{\beta}_{sc}) \tag{5}$$

Alternatively, the decomposition can also be done as

$$\ln \bar{Y}_{fc} - \ln \bar{Y}_{sc} = \sum \hat{\beta}_{sc}(\bar{X}_{fc} - \bar{X}_{sc}) + \sum \bar{X}_{fc}(\hat{\beta}_{fc} - \hat{\beta}_{sc}) \quad (6)$$

In equations (5) and (6) above, on the right-hand side, the first term can be interpreted as "endowment difference". The second term in these equations has been regarded in the literature as the "discrimination component". Studies use either of these alternative decomposition forms (Equations 5 or 6) based on their assumptions about the wage structure that would prevail in the absence of discrimination. This kind of problem is called the "index number problem".

Cotton, Neumark and Oaxaca/Ransom decomposition method

To solve the index number problem, Cotton (1988), Neumark (1988), and Oaxaca and Ransom (1994) propose an alternative decomposition method. The true non-discriminatory wage would lie somewhere between the FC and SC wage structure. The Cotton logarithmic wage differential is written as

$$\ln \bar{Y}_{fc} - \ln \bar{Y}_{sc} = \sum \beta^{*}(\bar{X}_{fc} - \bar{X}_{sc}) + \sum \bar{X}_{fc}(\hat{\beta}_{fc} - \beta^{*}) + \sum \bar{X}_{sc}(\beta^{*} - \hat{\beta}_{sc}) \quad (7)$$

The first term on the right-hand side of Equation (7) above refers to skill differences between SC and FC, while the second term represents the overpayment relative to FC due to favouritism and the third term refers to the underpayment to SC due to discrimination. The β^{*} is the reward structure that would have occurred in the absence of discrimination. The theory of discrimination provides some guidance in the choice of the non-discriminatory wage structure. In Cotton's (1988) decomposition, the assumption is operationalised by weighting the FC and SC wage structures by respective proportions of FC and SC in the labour force. The estimator β^{*} is defined as

$$\beta^{*} = P_{fc}\hat{\beta}_{fc} + P_{sc}\hat{\beta}_{sc} \quad (8)$$

where P_{fc} and P_{sc} are the sample proportions of FC and SC population; $\hat{\beta}_{fc}$ and $\hat{\beta}_{sc}$ are the FC and SC pay structures, respectively. The non-discriminatory or pooled wage structure proposed by Neumark (1988) and Oaxaca and Ransom (1994) is written below:

$$\beta^{*} = \Omega\hat{\beta}_{fc} + (I - \Omega)\hat{\beta}_{sc} \quad (9)$$

where I is the identity matrix, Ω is a weighting matrix, which is specified by

$$\Omega = (X'X)^{-1}(X'_{fc}X_{fc}) \tag{10}$$

where X is the observation matrix for the pooled sample, X_{fc} is the observation matrix for the FC sample. The interpretation of Ω as weighting matrix is readily seen by noting that

$$X'X = X'_{fc}X_{fc} + X'_{sc}X_{sc} \tag{11}$$

where X_{sc} is the observation matrix of the SC sample. Given $\hat{\beta}_{fc}$, $\hat{\beta}_{sc}$ and Equation (9), any assumption about β^* reduces to an assumption about Ω.

Machado–Mata–Melly decomposition method

This method was initially developed by Machado and Mata (2005). It is an extension of the Blinder–Oaxaca decomposition method, in the sense that instead of considering the difference at the mean of the wage distribution, it identifies the sources of wage gap at various quantiles of the wage distribution. The Machado–Mata decomposition is based on the estimation of marginal wage distributions consistent with a conditional distribution estimated using quantile regression. One can perform counterfactual exercises by comparing the marginal distributions implied by different distributions for the covariates. The counterfactual distributions are estimated as:

$$\left\{\ln \tilde{Y}_i^{cf} = x_i^{sc}\hat{\beta}_{u_i}^{fc}\right\}_{i=1}^{n} \text{ and } \left\{\ln \tilde{Y}_i^{cf} = x_i^{fc}\hat{\beta}_{u_i}^{sc}\right\}_{i=1}^{n}$$

The latest version of decomposition was developed by Melly (2006). The estimator of Melly decomposition will be numerically identical to the Machado–Mata decomposition if the number of simulations used in the Machado–Mata procedure goes to infinity. The mean square error in the Melly estimation is less than in the Machado–Mata estimation. The mean square errors of these two estimates converge only if the simulations in the Machado–Mata estimation become very large. The Melly estimator is also consistent and asymptomatically normally distributed. For the θth quantile, the wage gap between FCs and SCs can be decomposed into two components, as follows:

$$\hat{Q}_{fc}(\theta) - \hat{Q}_{sc}(\theta) = \underbrace{\left[\hat{Q}_{fc}(\theta) - \hat{Q}_{cf}(\theta)\right]}_{\text{Effect of Characterstics}} + \underbrace{\left[\hat{Q}_{cf}(\theta) - \hat{Q}_{sc}(\theta)\right]}_{\text{Effect of Coefficients}} \tag{12}$$

where $\hat{Q}_{fc}(\theta)-\hat{Q}_{sc}(\theta)$ is the wage gap estimated from the θth quantile of the unconditional log wage distribution for FCs and SCs respectively and $\hat{Q}_{cf}(\theta)$ is the estimated counterfactual unconditional quantile of the log wage distribution for SCs created using the coefficients of FCs. It represents the wage distribution of SCs that would have prevailed if SCs had been endowed with their own characteristics but get paid like FCs.

The first component on the right hand side of Equation (12) can be interpreted as characteristics effect or endowment differences. The second term has been regarded in the literature as the coefficients effect or discrimination component.

Empirical results and discussion

The following descriptive analyses have been carried out to study the wage structure by social groups and age cohorts. Subsequently, an attempt has been made to measure the extent of caste discrimination by age cohorts in the labour market. In the light of the adopted "reservation policy" aimed at protecting the Scheduled Caste population, we have divided the sample into two demographic components (young cohorts: population age between 18 and 40 years and old cohorts: population age between 41 and 65 years).

Wage differentials by social groups in the labour market: young vs old cohorts

In this section, we focus only on regular workers as the caste-based wage disparities are found to be negligible among casual workers. The average daily wages and wage ratios of regular workers by social groups in rural and urban areas are reported in Table 8.1. We find that the wages of SCs are lower than that of forward castes (FCs), irrespective of age cohorts. The SC to forward caste wage ratios are found to be lower in urban than in rural areas. Thus, the caste-based wage differential seems to be more in urban than in rural areas irrespective of age cohorts. The caste-based wage gap is found to be declining over the years from 2005 to 2012 irrespective of age cohorts, except for old workers in urban areas.

A further disaggregation of wage differentials by levels of education is given in Table 8.2. We find that in the regular urban labour market, caste-based wage differential among young workers is the highest at graduate and above level; in 2012, the wages paid to SC graduates are about 71 per cent that of forward castes. The SC to FC wage ratio at this level of education has declined from 0.81 in 2005 to 0.71 in 2012. This implies that caste-based wage gap attributed to graduation and above levels of education is increasing over the years in urban India. Similarly, for old cohorts,

Table 8.1 Average real daily wages of regular workers by social groups and age cohorts, 1983 to 2012 (2001–2002 prices)

	1983				*1994*				*2005*				*2012*			
	SC	*FC*	*t-stat of MD*	*Wage Ratio (SC/FC)*	*SC*	*FC*	*t-stat of MD*	*Wage Ratio (SC/FC)*	*SC*	*FC*	*t-stat of MD*	*Wage Ratio (SC/FC)*	*SC*	*FC*	*t-stat of MD*	*Wage Ratio (SC/FC)*
Young Cohort (18–40 years)																
Rural	39.12	56.39	11.41*	0.69	61.52	77.24	6.91*	0.80	83.57	127.05	3.78*	0.66	103.48	135.26	4.39*	0.76
Urban	68.94	90.68	14.59*	0.76	92.22	116.68	11.60*	0.79	94.25	158.54	15.48*	0.59	137.75	220.39	11.89*	0.63
Total	52.65	77.52	22.62*	0.68	77.14	102.73	15.72*	0.75	89.04	148.94	13.58*	0.60	123.39	194.79	13.61*	0.63
Old Cohort (41–65 years)																
Rural	41.29	69.60	12.87*	0.59	82.66	119.20	9.21*	0.69	138.54	205.41	8.49*	0.67	178.61	231.86	3.99*	0.77
Urban	79.71	121.94	15.79*	0.65	127.93	177.53	12.03*	0.72	180.15	266.86	11.15*	0.68	216.51	328.74	8.19*	0.66
Total	60.19	103.11	21.93*	0.58	106.68	156.74	16.49*	0.68	161.78	247.93	14.87*	0.65	201.09	299.98	9.75*	0.67

Source: Computed from unit level data of various NSS rounds.

Note
1 FC includes OBCs in 1983 and 1993–1994.
2 MD (FC-SC) denotes Mean Difference between Forward Castes (FC) and Scheduled Castes (SC).
3 * denote significant at 1 per cent level.

Table 8.2 Average real daily wage (in Rs.) of regular urban workers by social groups, levels of education and age cohorts, 2005 to 2012 (2001–2002 prices)

Educational Levels	*2005*			*2012*		
	SC	*FC*	*SC/FC*	*SC*	*FC*	*SC/FC*
Young Cohort (18–40 years)						
Illiterate	61.64	73.40	0.84	62.73	79.00	0.79
Literate and up to Primary	67.16	82.32	0.82	89.91	92.76	0.97
Middle	79.81	87.97	0.91	92.88	105.17	0.88
Secondary and Higher Secondary	99.58	124.85	0.80	127.77	144.14	0.89
Diploma/ Certificate	147.51	222.05	0.66	200.27	225.62	0.89
Graduate and above	200.58	247.85	0.81	247.69	347.81	0.71
Total	94.27	158.56	0.59	137.75	220.39	0.63
Old Cohort (41–65 years)						
Illiterate	87.33	79.91	1.09	104.01	87.26	1.19
Literate and up to Primary	136.10	117.80	1.16	132.50	134.15	0.99
Middle	166.15	160.94	1.03	168.94	166.89	1.01
Secondary and Higher Secondary	260.58	238.61	1.09	270.35	262.76	1.03
Diploma/ Certificate	346.03	309.31	1.12	318.85	393.61	0.81
Graduate and above	314.13	383.77	0.82	387.91	480.29	0.81
Total	180.15	266.86	0.68	216.51	328.74	0.66

Source: Author's calculation.

Note: Sample of individuals belongs to 15–65 age groups in 18 major states.

we observe a significant wage gap between SC and FC urban workers at diploma, graduation and above degrees.

Empirical evidence on caste discrimination in the Indian labour market: young vs old age cohorts

In this section, we restrict our sample to regular urban workers as the caste-based wage differential is found to be higher in this segment of the labour market. We have tried to find out whether this wage gap between SCs and FCs is on account of human capital difference or discrimination in the labour market. For this purpose, we have applied different approaches of the decomposition method. We prefer to analyse caste discrimination against young (18–40 years) and old workers (41–65 years) to study the effectiveness of the reservation policy, which is introduced by the Government of India to improve the participation in the labour force for deprived and privileged classes by way of reserving seats in various public places, viz., in opportunity in jobs, education, and other fields of life. However, the reservation policy is not directed to different age groups.

As discussed in our methodology section, initially the single-equation method is adopted, in which we have added caste dummy variable in the wage equation. Due to space constraints, all the econometric results have not been presented here. We find that in the regular urban labour market, SC young employees earn 9 per cent less than Forward Caste young employees in 1994, 19 per cent less in 2005 and 17 per cent less in 2012, while SC old employees earn 6 per cent less than Forward Caste (FC) employees in 1994, 7 per cent less in 2005 and 13 per cent less in 2012. Thus, there exists a wage gap between SC and Forward Caste workers irrespective of age cohorts and for old workers, it is growing over the years. One of the limitations with the above single-equation-based analysis is that it assumes slope coefficients to be the same for both SC and Forward Caste equations. Thus, our subsequent analysis estimates earnings function separately for SCs and FCs and subjected to the earnings equation to Blinder–Oaxaca decomposition, where we decompose the gross wage gap into two components, such as endowment difference and discrimination in the labour market.

Table 8.3 reports the wage regression estimates for regular urban workers by age cohorts and social groups. The natural logarithm of real daily wage rate is used as the dependent variable; and age, level of education, gender, marital status, sector, nature of employment and regions are taken as the explanatory variables. The descriptive statistics of the variable used in the regression analysis are given in appendix Table A8.1. The age variable is taken as the proxy of experience. We find that both age and education variables are statistically significant irrespective of age cohorts and social groups. The sign of coefficients of age variable is positive and age square is negative. This implies that there is an increment in earnings with age and it starts declining after the retirement age. As expected, for both young and old cohorts, the value of education coefficients is positive and it is increasing with levels of education irrespective of social groups. Interestingly, for the young age cohort, the value of education coefficients is higher for SCs than that of FCs at middle, higher secondary and diploma education, while for the old cohort, the value of education coefficients is higher for FCs than that of SCs for all levels of education. Moreover, for both the young and old cohorts, the gender variable is found to be significant irrespective of social groups, but the value of the male dummy coefficient is higher for SCs than Forward Castes. This implies that the gender wage gap is higher among SCs

Table 8.3 Estimates of augmented earnings equation for regular urban workers by social groups: young vs old cohorts, 2012

Variables	*Young Cohort (18–40 years)*				*Old Cohort (41–65 years)*			
	Forward Castes		*SC*		*Forward Castes*		*SC*	
	Coeff.	*t-stats*	*Coeff.*	*t-stats*	*Coeff.*	*t-stats*	*Coeff.*	*t-stats*
Age	0.05	3.24	0.05	2.38	0.20	5.28	0.30	5.23
Age Sq	–0.001	–2.24	–0.001	–1.49	–0.002	–5.13	–0.003	–5.17
Primary	0.12	2.78	0.12	2.2	0.21	3.38	0.19	2.72
Middle	0.19	4.87	0.20	4.34	0.34	6.17	0.28	4.36
Secondary	0.35	9.2	0.35	7	0.57	10.98	0.55	8.07
HSC	0.50	12.06	0.52	10	0.72	12.95	0.64	8.6
Diploma	0.82	15.37	0.88	11.71	0.95	15.24	0.87	9.42
Grad and above	1.08	29.16	0.92	19.65	1.10	22.23	0.89	12.31
Male	0.37	14.19	0.44	13.31	0.28	7.95	0.49	6.9
Married	0.11	4.41	0.01	0.4	0.23	4.59	–0.02	–0.27
Public	0.38	14.57	0.45	13.57	0.62	22.21	0.84	17.29
Permanent	0.26	11.98	0.27	8.93	0.33	7.38	0.26	4.02
South	0.04	1.5	0.07	2.1	–0.02	–0.45	–0.03	–0.64
East	–0.16	–5.52	–0.08	–2.01	–0.05	–1.88	–0.01	–0.17
West	0.16	6.93	0.07	1.79	0.05	1.76	0.02	0.33
constant	2.64	10.98	2.47	7.74	–1.35	–1.46	–3.94	–2.77
R squared	0.43		0.48		0.51		0.62	
Number of observations	5009		1905		3325		1018	

Source: Author's calculation.

Note: Dependent variable is the natural logarithm of real daily wage.

than that of FCs. Besides, marriage has a positive and significant effect on the earnings of Forward Castes but not for SCs. The coefficient of the public sector dummy is positive and significant for both SCs and FCs; the value of the coefficient of the public sector dummy is higher for SCs than FCs. This implies that SCs are in a better position in the public sector of the regular urban labour market. Workers in permanent jobs are getting better earnings than their counterparts irrespective of social groups. For the young age cohort, being in the eastern region negatively affects the earnings of FCs and SCs, while being in the western region positively affects the earnings of FCs and being in the southern region positively affects the earnings of SCs. But for the old age cohort, region variables are found to be insignificant.

Our subsequent discussion focuses on analysing the recent trend of caste discrimination in the regular urban labour market. For observing the historical trends, we have examined the issue of caste discrimination by age cohorts over time from 1983 to 2012. We have used different approaches of the decomposition method for measuring the extent of caste discrimination in the regular urban labour market.

Oaxaca–Blinder decomposition results

We have used the Oaxaca–Blinder decomposition method to decompose the wage gap between FC and SC workers into endowment difference and discrimination in the regular urban labour market; a separate analysis has been done for young and old workers. The results are given in Table 8.4. We find that, for both young and old age cohorts, the contribution of endowment difference to raw wage gap between SC and FC workers is larger than that of discrimination. This suggests that pre-labour market discriminatory practices against SCs in access to education and skill development are more crucial in explaining wage differentials than labour market discrimination. Besides, the discrimination effect is higher for the young cohort than old cohort irrespective of the study period. This finding is consistent with the Arabsheibani et al. (2012) study. From our results, one may also perceive that the government intervention programmes to improve the social connectedness of SCs with the privileged class have been less effective during the study period. However, for the young cohort, the caste-based wage gap as well as discrimination at the mean of the wage distribution has declined over the years from 2005 to 2012.

The data on OBCs is not available for the year 1983 and 1994. So, for comparison purpose, Figures 8.1 and 8.2 depict the results of the decomposition

Table 8.4 Results of Oaxaca–Blinder decomposition – SCs vs FCs – among young and old cohorts of regular urban workers, 1983 to 2012

Components	*1983**	*1994**	*2005*	*2012*
Young (Age Cohort: 18–40 Years)				
Raw wage Differentials	0.27	0.21	0.42	0.40
Explained (endowment)	0.24	0.13	0.23	0.24
Unexplained (Discrimination)	0.03	0.08	0.19	0.16
Endowment Difference (%)	90.21	61.17	53.96	60.99
Discrimination (%)	9.79	38.83	46.04	39.01
Old (Age Cohort: 41–65 Years)				
Raw wage Differentials	0.37	0.32	0.42	0.42
Explained (endowment)	0.37	0.27	0.36	0.30
Unexplained (Discrimination)	0.003	0.05	0.06	0.12
Endowment Difference (%)	99.17	83.61	85.34	71.12
Discrimination (%)	0.83	16.39	14.66	28.88

Source: Computed from NSS data.

Note: *denotes FCs include OBCs for 1983 and 1993–94.

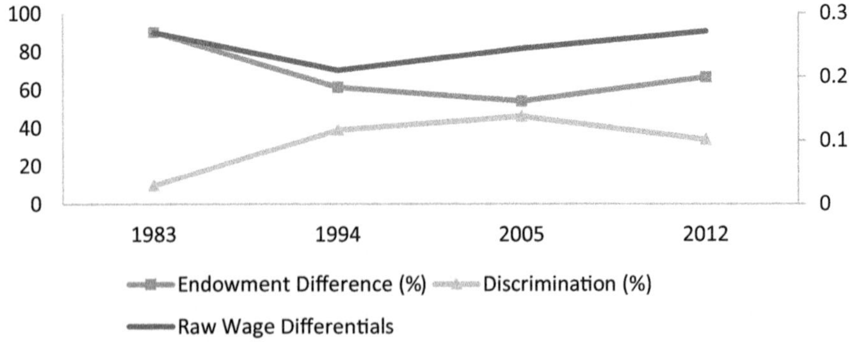

Figure 8.1 Decomposition of wage gap between SC and Other (OBC and FC) workers belonging to young cohorts in regular urban LM, 1983–2012

Source: Estimated from different rounds of the NSSO's survey data.

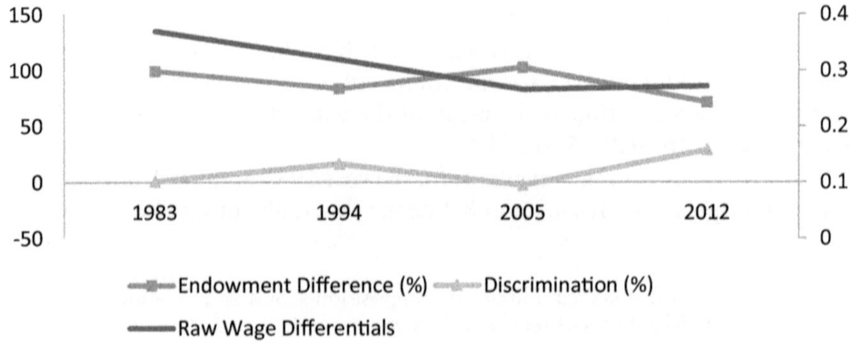

Figure 8.2 Decomposition of wage gap between SC and Other (OBC and FC) workers belonging to old cohorts in regular urban LM, 1983–2012

Source: Estimated from different rounds of the NSSO's survey data.

of the wage gap between SCs and others (OBCs and FCs) by age cohorts, over time. We find that for the young cohort, the wage gap has increased from 0.21 in 1994 to 0.27 in 2012. This is largely attributed to increasing endowment difference between SCs and others in the regular urban labour market. On the other hand, for the old cohort, the wage gap between SCs and others (OBCs and FCs) has declined from 0.32 in 1994 to 0.27 in 2012. This is largely attributed to the declining endowment difference between SCs and others in the regular urban labour market.

Further, we have measured the extent of caste discrimination in public and private sectors separately in order to study the effectiveness of reservation policy. As expected, the extent of discrimination against SCs is more in the private sector than in the public sector, irrespective of age cohorts. In addition to this, the extent of discrimination against young workers is more than that for old workers, irrespective of the sector of employment. However, in both public and private sectors, the extent of discrimination against young workers is declining over the years from 2005 to 2012 (See Table 8.5).

Subsequently, we have compared different approaches of the decomposition method and find that the Oaxaca and Ransom method has the smallest standard error and should probably be preferred over other approaches (see Table 8.6). Using the Oaxaca and Ransom decomposition method, we find that the wage gap between SCs and FCs is largely attributable to the endowment difference and the effect of characteristics is more for old workers than young workers. In 2012, the contribution of discrimination to raw wage differentials between SCs and FCs is 37.5 per cent for young workers, while it is 27.7 per cent for old workers. We have further decomposed the discrimination component into overpayment to FCs and underpayment to SCs in the labour market. We find that the underpayment or treatment disadvantage to SCs (cost of being SC in the labour market) is more for the young cohort than the old cohort. Overall, we find that the cost of being SCs in the labour market is very high. They are hugely underpaid in the labour market.

However, the reliability of the mean-based decomposition analysis has been questioned in various empirical studies. It is argued that these methods are not

Table 8.5 Oaxaca–Blinder decomposition results for young and old workers in public and private sectors of regular urban LM, SCs vs FCs

Components	*1994**		*2005*		*2012*	
	Public Sector	*Private Sector*	*Public Sector*	*Private Sector*	*Public Sector*	*Private Sector*
Young (Age Cohort: 18–40 Years)						
Raw Wage Differentials	0.29	0.34	0.44	0.49	0.41	0.44
Endowment Difference (%)	86.5	63.2	62.5	60.0	71.5	66.6
Discrimination (%)	13.5	36.8	37.5	40.0	28.5	33.4
Old (Age Cohort: 41–65 Years)						
Raw Wage Differentials	0.33	0.46	0.36	0.63	0.32	0.62
Endowment Difference (%)	80.0	86.6	72.6	89.5	75.2	72.2
Discrimination (%)	20.0	13.4	27.4	10.5	24.8	27.8

Source: Author's calculation.

Note: *denotes FCs include OBCs for 1994.

Table 8.6 Reimers–Cotton–Neumark – Oaxaca and Ransom approach – SCs vs FCs (young and old workers in regular urban LM), 2012

Components	*Reimers / Cotton (w=0.5)*	*Neumark/ Oaxaca and Ransom (w = omega)*	*Oaxaca–Blinder Using Male means as weight (w=1)*	*Oaxaca–Blinder Using Female means as weight (w=0)*
Young (Age Cohort: 18–40 Years)				
Raw Wage Differential	**0.40 (0.0225)**			
Endowment Difference (%)	55.2 (0.0165)	62.5 (0.0168)	61.0 (0.0173)	49.4 (0.0179)
Discrimination (%)	44.8 (0.0177)	37.5 (0.0159)	39.0 (0.0182)	50.6 (0.0193)
Overpayment to male (%)	25.3 (0.0096)	10.3 (0.0044)	-	-
Underpayment to female (%)	19.5 (0.0091)	27.2 (0.0114)	-	-
Old (Age Cohort: 41–65 Years)				
Raw Wage Differential	**0.42 (0.0344)**			
Endowment Difference (%)	63.9 (0.0274)	72.3 (0.0269)	71.1 (0.0270)	56.7 (0.0312)
Discrimination (%)	36.1 (0.0241)	27.7 (0.0211)	28.9 (0.0247)	43.3 (0.0275)
Overpayment to male (%)	21.7 (0.0137)	6.5 (0.0050)	-	-
Underpayment to female (%)	14.4 (0.0123)	21.2 (0.0162)	-	-

Source: Computed from NSS data.

Note:
(1) Unexplained Component = overpayment + underpayment component.
(2) Figures in parentheses indicate standard errors.

appropriate to measure the persistent effect of caste discrimination across the wage distribution. Thus, subsequently, we examine the effect of caste discrimination across the wage distribution by using a quantile-based decomposition method proposed by Machado and Mata (2005) and Melly (2006).

Wage gap across quantiles of the wage distribution

From Table 8.7, we find that in the regular urban labour market, the extent of caste-based wage gap as well as discrimination varies significantly across the

Table 8.7 MMM decomposition results across quantiles: FCs vs SCs – among young and old cohorts of regular urban workers, 2005 and 2012

Components	*2005*						*2012*					
	10th	*25th*	*50th*	*75th*	*90th*	*OLS*	*10th*	*25th*	*50th*	*75th*	*90th*	*OLS*
Young (Age Cohort: 18–40 Years)												
Raw Wage Differentials	0.41	0.38	0.42	0.46	0.44	0.42	0.31	0.31	0.39	0.49	0.49	0.40
Characteristics effect (endowment difference)	0.23 (56.9)	0.20 (53.3)	0.20 (47.9)	0.21 (46.4)	0.19 (44.3)	0.23 (54.0)	0.20 (63.4)	0.19 (60.3)	0.21 (52.6)	0.20 (41.3)	0.17 (35.1)	0.24 (61.0)
Coefficients (Discrimination)	0.18 (43.1)	0.18 (46.7)	0.22 (52.1)	0.24 (53.6)	0.24 (55.7)	0.19 (46.0)	0.12 (36.6)	0.13 (39.7)	0.18 (47.4)	0.29 (58.7)	0.32 (64.9)	0.16 (39.0)
Old (Age Cohort: 41–65 Years)												
Raw Wage Differentials	0.48	0.53	0.42	0.37	0.32	0.42	0.46	0.53	0.46	0.34	0.26	0.42
Characteristics effect (endowment difference)	0.37 (77.3)	0.34 (63.9)	0.28 (67.3)	0.23 (62.3)	0.14 (41.8)	0.36 (85.3)	0.30 (65.0)	0.27 (51.4)	0.24 (52.4)	0.18 (51.6)	0.16 (59.6)	0.30 (71.1)
Coefficients (Discrimination)	0.11 (22.7)	0.19 (36.1)	0.14 (32.7)	0.14 (37.7)	0.19 (58.2)	0.06 (14.7)	0.16 (35.0)	0.26 (48.6)	0.22 (47.6)	0.17 (48.4)	0.11 (40.4)	0.12 (28.9)

Source: Calculated from NSS unit level data.

Note: Percentage share of raw wage difference are given in parenthesis.

quantiles of the wage distribution irrespective of age cohorts. For the young cohort, the wage gap attributable to discrimination is higher at the top wage quantiles than at bottom wage quantiles. The pattern remains the same for the year 2005 and 2012. But a reverse pattern is observed for the year 1994, in which the discrimination effect is found higher at the bottom wage quantiles than at top wage quantiles. During 2005 to 2012, the contribution of discrimination to the wage gap between SCs and FCs is rising over the years for top wage quantiles, such as 75th and 90th quantiles. Using Arulampalam, Booth, and Bryan's (2007) criteria, we find that there exists a "glass ceiling effect", particularly for workers belonging to young age cohorts as the 90th percentile wage gap is higher than the estimated wage gaps in other parts of the wage distribution. The glass ceiling effect primarily reflects the fact that SCs and Forward Castes have different jobs or there exists occupational segregation in the labour market. Comparing the effect of discrimination on the wage gap between SCs and Forward Castes for the young and old cohorts, we find that in 2012, the young cohort at the top of the distribution are more discriminated against compared to the old cohort (see Figure 8.3), while for the year 1994 and 2005, the effect of discrimination is found higher for the young cohort than the old cohort irrespective of wage quantiles, excepting 90th quantile in 2005.

Further, we have measured the extent of caste discrimination in public and private sectors across the wage quantiles (see Table 8.8). As expected, the extent of discrimination against SCs is more in the private sector than in the public sector, irrespective of age cohorts and wage quantiles. In addition to this, the extent of discrimination against young workers is more than that for old workers, irrespective of the sector of employment and wage quantiles. We observe the "glass ceiling effect" for young workers in the private sector as the extent of caste-based wage gap as well as discrimination is increasing across the wage quantiles (see Figure 8.4), while the "sticky

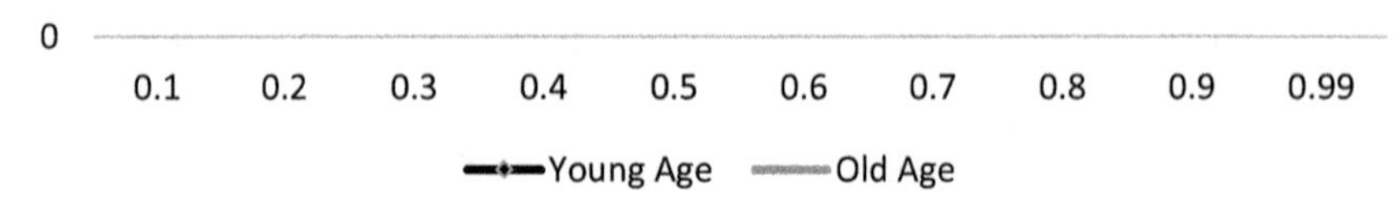

Figure 8.3 Caste discrimination coefficient across wage quantiles between young and old cohorts in regular urban LM, 2012

Source: Based on the results presented in Table 8.7.

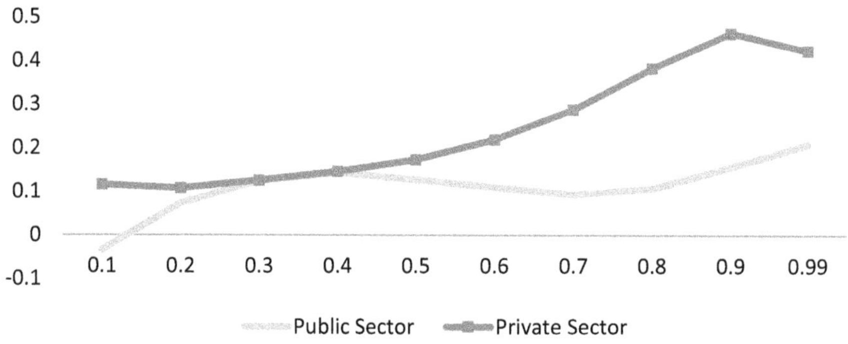

Figure 8.4 Caste discrimination coefficients across wage quantiles in public and private sectors for regular urban young (18–40 years) workers, 2012

Source: Calculated from NSS unit level data

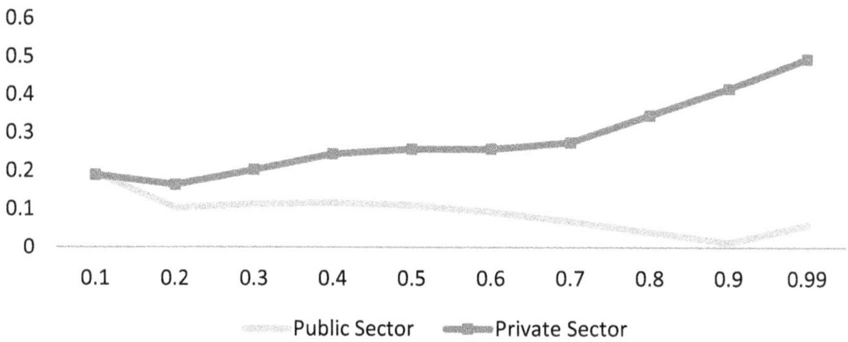

Figure 8.5 Caste discrimination coefficients across wage quantiles in public and private sectors for regular urban old (41–65 years) workers, 2012

Source: Calculated from NSS unit level data.

floor effect" is observed for old workers in the public sector as the extent of caste-based wage gap as well as discrimination is declining across the wage quantiles (see Figure 8.5).

Conclusions

Faster growth without social stability is not sustainable. Although India's recent growth success story has encouraged academics and policy makers alike to predict faster speed of convergence of per capita income to developed countries within a time less than about three decades, the observed and rising social and economic differential in development within this success story has been a constant worry. Under this backdrop, this chapter has investigated how discrimination across demographic cohorts and social groups is responding to the growth spurts. For observing the historical trends, the study examines the issue of caste discrimination by age cohorts over time using different rounds of NSSO's survey data spanning the period 1983 to 2011–2012. The role and effect of caste discrimination between young and old age cohorts have been emphasised in the study. It is to be noted that both the population cohorts have their own social affiliation which has been endogenously determined within the Indian context, viz., whether they belong to high or low caste. Comparing the effect of discrimination between young and old cohorts for the same social group, we find that the discrimination effect is fairly acute among the young population. Besides, for the young cohort, the wage gap between SCs and FCs attributable to discrimination is higher at the top wage quantiles than at bottom wage quantiles, while at the top wage quantiles, the contribution of discrimination to the wage gap is rising over the years from 2005 to 2012. Moreover, the extent of caste discrimination is more in the private sector than in the public sector, irrespective of age cohorts and wage quantiles. In addition to this, young workers are more discriminated against than old workers, irrespective of the sector of employment and wage quantiles. We observe the "glass ceiling effect" among young workers in the private sector and "sticky floor effect" among old workers in the public sector. It is to be noted that in the population cohort, if discrimination is more persistent in the young than the old, the future value of economic growth is negatively affected through social tension and uncertainty.

In general, our results suggest that the effect of reservation systems has no perceived impact on young cohorts, particularly in the private sector. Thus, there is need for extending the affirmative action programme to the private sector. The demand for reservation in the private sector can be justified on the ground that the Government of India provides various incentives and safeguards to the private sector units to promote their business, thereby creating better situations for the encouragement of business and trade. Further, the foreign policy and export-import policy of the government of India contribute to the betterment of business, set up by individuals in the private sector. Foreign investors also invest in the private sector by purchasing shares, which is made possible by the policies of the government. The private sector, in turn, is expected to fulfil its social responsibility. The upliftment of the weaker sections of society is a stated objective of our country, and thus the implementation of reservation in the private sector is part of the social responsibility of both the government and the private sector. In fact, it is merely the fulfilment of the constitutional agenda of distributive justice enshrined in various articles and clauses of the Constitution. If, however, the private sector is not fulfilling its social responsibility, the government should make such provisions through legislative measures.

Our empirical evidence shows that a large part of the wage gap between SCs and Forward Castes is explained by the differences in human capital endowments, that is the education, skills and experience of a worker. So, there is a need for effective measures to improve the educational and skill endowment of workers belonging to the disadvantaged social group. These measures may include not only support for education and skill formation, but also steps towards alleviation of their poverty. Their access to education, training and health facilities, as also sources of livelihood, needs to be improved. For, while quite often, the availability of institutions and the facility is not found sufficient, enabling people to avail of these is also equally important, insofar as their existing economic and social handicaps prevent their use (Papola, 2012).

It is, however, observed that endowment again is a necessary but not a sufficient condition when it comes to benefitting from participation in the labour market and accessing jobs befitting one's qualifications. Part of this exclusion takes place because the dissemination of information on jobs is often exclusionary: information becomes available only to those who have someone "inside"; and the insiders mostly happen to be from among those socially and economically better placed. In the process, equality of opportunity is denied. The discriminatory process can extend beyond access to information to processes of selection where attributes that have little relevance to the performance of a given job tend to favour candidates with better social and economic endowment (e.g. felicity with spoken English). The second necessary condition for reducing exclusion and discrimination, therefore, is to ensure equality of opportunity in access to information and the use of non-discriminatory methods and criteria in selection (Papola, 2012).

Notes

1 The "glass ceiling effect" is said to occur if caste-based wage gaps increase throughout the wage distribution and accelerates in the upper tail (Albrecht, Bjorklund, & Vroman, 2003).

2 The "sticky floor effect" is said to occur if caste-based wage gaps are higher at lower ends of the distribution and steadily decline over the distribution (Deshpande, Goel, & Khanna, 2015).

References

Akerlof, A. G. (1976). The economics of caste and of the rat race and other woeful tales. *The Quarterly Journal of Economics*, *90*(4), 599–617.

Akerlof, A. G. (1980). The theory of social customs, of which unemployment may be one consequence. *The Quarterly Journal of Economics*, *94*(4), 749–775.

Albrecht, J., Bjorklund, A., & Vroman, S. (2003). Is there a glass ceiling in Sweden? *Journal of Labor Economics*, *21*(1), 145–177.

Ambedkar, B. R. (1936). *Annihilation of caste*. Jalandhar: Bhim Patrika.

Ambedkar, B. R. (1987a). The Hindu social order: Its essential features. In V. Moon (Ed.), *Dr. Babasaheb Ambedkar writings and speeches* (Vol. 3, pp. 96–115). Maharashtra: Education Department, Govt of Maharashtra.

Ambedkar, B. R. (1987b). The Hindu social order: Its unique features. In V. Moon (Ed.), *Dr. Babasaheb Ambedkar writings and speeches* (Vol. 3, pp. 116–129). Maharashtra: Education Department, Govt of Maharashtra.

Arabsheibani, G., Gupta, P., & Mishra, T. (2012). *Youth, endogenous discrimination and development conundrum in India*. Retrieved from http://conference.iza.org/conference_files/worldb2012/gupta_p8290.pdf (accessed 24 May 2018).

Arrow, K. J. (1972). Some mathematical models of job discrimination in the labour market. In A. H. Pascal (Ed.), *Racial discrimination in economic life*. Lexington, MA: Lexington Press.

Arrow, K. J. (1973). The theory of discrimination. In O. Ashenfelter & A. Rees (Eds.), *Discrimination in labour markets* (pp. 3–33). Princeton, NJ: Princeton University Press.

Arulampalam, W., Booth, A. L., & Bryan, M. L. (2007). Is there a glass ceiling over Europe? Exploring the gender pay gap across the wage distribution. *Industrial and Labour Relations Review*, *60*(2), 163–186.

Azam, M. (2012). A distributional analysis of social group inequality in rural India. *Journal of International Development*, *24*(4), 415–432.

Becker, G. S. (1971). *The economics of discrimination* (2nd ed.). Chicago: University of Chicago Press.

Birdsall, N., & Sabot, R. (1991). *Unfair advantage: Labour market discrimination in developing countries*. Regional Sectoral Studies, Washington, DC: The World Bank.

Blinder, A. S. (1973). Wage discrimination: Reduced form and structural estimates. *Journal of Human Resources*, *8*(4), 436–455.

Cotton, J. (1988). On the decomposition of wage differentials. *The Review of Economics and Statistics*, *70*(2), 236–243.

Darity, W. (1997). Reparations. In S. Myers (Ed.), *Civil rights and race relations in the post-Reagan Bush era*. London: Praeger.

Deshpande, A., Goel, D., & Khanna, S. (2015). *Bad Karma or discrimination? Male-female wage gaps among salaried workers in India* (Working Paper No. 243.) New Delhi: Centre for Development Economics, Delhi School of Economics.

Hann, De A. (1997). Poverty and social exclusion: A comparison of debates on deprivation (Working Paper No.2, Poverty Research Unit at Sussex). Brighton: University of Sussex.

Kijima, Y. (2006). Caste and tribe inequality: Evidence from India, 1983–1999. *Economic Development and Cultural Change*, *54*(2), 369–404.

Labour Bureau. (various years). *Indian labour statistics*. Government of India.

Lal, D. (1988). *Hindu equilibrium: Cultural stability and economic stagnation* (Vol. I). Oxford: Clarendon Press.

Machado, J. A. F., & Mata, J. (2005). Counterfactual decomposition of changes in wage distributions using quantile regression. *Journal of Applied Econometrics*, *20*(4), 445–465.

Madden, J. F. (1975). *The economics of sex discrimination*. Lexington, MA: D.C Health & Co.

Melly, B. (2006). *Estimation of counterfactual distributions using quantile regression* (Discussion Paper). St. Gallen: University of St. Gallen.
Mincer, J. (1974). *Schooling, experience, and earnings* (Human Behavior & Social Institutions No. 2). New York: Columbia University Press.
Neumark, D. (1988). Employers' discriminatory behaviour and the estimation of wage discrimination. *Journal of Human Resources*, *23*(3), 279–295.
NSSO. (2014). *Employment and unemployment situation in India* (68th Round, July 2011-June 2012, Report No. 554). New Delhi: National Sample Survey Organization.
Oaxaca, R. L. (1973). Male-female wage differentials in urban labour markets. *International Economic Review*, *14*(3), 693–709.
Oaxaca, R. L., & Ransom, M. R. (1994). On discrimination and the decomposition of wage differentials. *Journal of Econometrics*, *61*(1), 5–21.
Papola. (2012). *Social exclusion and discrimination in the labour market* (Working Paper No. 2012/04.) New Delhi: Institute for Studies in Industrial Development.
Phelps, E. S. (1972). The statistical theory of racism and sexism. *American Economic Review*, *62*(4), 659–661.
Scoville, J. G. L. (1991). Towards a model of caste economy. In J. G. Scoville (Ed.), *Status influences in third world labour markets: Caste, gender and custom* (pp. 386–393). New York: Walter de Gruyter.
Sen, A. (2000). *Social exclusion: Concept, application and scrutiny* (Working Paper, Social Development Paper No. 1, June). Bangkok: Asian Development Bank.
Tzannatos, Z. (1987). A general equilibrium model of discrimination and its effects on income. *Scottish Journal of Political Economy*, *34*(1), 19–36.

APPENDIX

Table A8.1 Descriptive statistics of variables used in the augmented earnings equation for regular urban workers by social group: young vs old cohorts, 2012

Variables	*Description of Variables*	*Young Age Cohorts*				*Old Age Cohorts*			
		FC		*SC*		*FC*		*SC*	
		Mean	*S.D.*	*Mean*	*S.D.*	*Mean*	*S.D.*	*Mean*	*S.D.*
Ln Real Wage	Natural Logarithm of real daily wage at 2001 prices (in Rupees)	4.97	0.89	4.56	0.82	5.49	0.90	5.06	0.97
Age	Age in Years	30.38	6.10	29.94	6.46	49.52	5.65	49.31	5.59
Age Sq	Age Square (in years)	960.15	368.99	938.05	387.22	2484.03	575.48	2462.75	568.51
Primary	If the worker has completed primary education =1;0 otherwise	0.07	0.25	0.11	0.32	0.05	0.22	0.11	0.31
Middle	If the worker has completed middle education =1;0 otherwise	0.13	0.33	0.19	0.39	0.09	0.29	0.15	0.36
Secondary	If the worker has completed secondary education=1;0 otherwise	0.13	0.34	0.14	0.35	0.16	0.37	0.15	0.36
HSC	If the worker has completed higher secondary education=1;0 otherwise	0.13	0.34	0.12	0.33	0.12	0.32	0.10	0.30
Diploma	If the worker has completed diploma or certificate course =1; 0 otherwise	0.05	0.21	0.04	0.20	0.04	0.21	0.03	0.18
Grad and above	If the worker has completed graduate and above degree education=1;0 otherwise	0.42	0.49	0.21	0.40	0.45	0.50	0.19	0.39
Male	If the individual is a male =1; 0 otherwise	0.78	0.41	0.74	0.44	0.81	0.39	0.76	0.43
Married	If the individual is currently married =1; 0 otherwise	0.64	0.48	0.65	0.48	0.91	0.29	0.86	0.34
Public Sector	If the individual is working in public sector=1; 0 otherwise	0.24	0.43	0.28	0.45	0.55	0.50	0.58	0.49
Permanent	If the worker is having a permanent job =1; 0 otherwise	0.68	0.46	0.65	0.48	0.88	0.32	0.85	0.36
South	If the individual working in South =1; 0 otherwise	0.16	0.37	0.29	0.45	0.17	0.37	0.26	0.44
East	If the individual is working in East and North East region=1; 0 otherwise	0.17	0.38	0.18	0.39	0.24	0.43	0.24	0.42
West	If the individual is working in West =1; 0 otherwise	0.28	0.45	0.20	0.40	0.21	0.41	0.19	0.39

Note: S.D. denotes Standard Deviation.

9

MULTIPLE MARGINALITIES

Educational and occupational mobility of differently abled persons across social groups in India

C. M. Lakshmana

Introduction

Although disability studies have been an important part of sociology and development literature, adequate attention has not been paid to its interlocking relationship with class, caste and other markers of social stratification (Jenkins, 1991). By drawing from secondary data sources, this chapter attempts to situate "disability" in the wider canvass of social stratification, especially the caste division in Indian society and attempts to understand the upward mobility of persons with disability across social groups in India. We examine the status of disability in India in general and across social groups in particular. To be more specific, we attempt to analyse the major change of disabled persons to the total population of India, in terms of percentage as well as the growth rates, the level of literacy and educational status of disabled persons in India by residence and sex for the last one decade, the gaps in literacy status and educational level with a comparative picture across social groups by sex and residence between 2001 and 2011.

As revealed by the World Health Organisation (WHO), currently around 400 million disabled persons live in the developing world and a majority of them are poor. And poverty is the most important cause of their disability (Groce, Kett, Lang, & Trani, 2011). It is estimated that at least 10 per cent of the developing world's population is disabled in one way or the other (Groce et al., 2011). The disabled are deprived of all opportunities for social and economic development. Chaudhuri (2006) noted that basic facilities like health, education and employment are denied to this vulnerable section of the population. It is estimated that about 40 million of the more than 100 million out-of-school children have disabilities and, around 70 per cent of the disabled are unemployed (Chaudhuri, 2006). The disabled are also not a homogeneous group and there are different types of disabilities, with different requirements, and the problems as well as attention and help required by each type are different from the others (Chaudhuri, 2006).

As per the 2011 Census, the total population of India is 1.23 billion, of which about 2.21 per cent (26.81 million people) suffer from disability of one kind or the other. Of the total disabled population, males and females account for 14.98 and 11.82 million respectively. As regards the rural/urban divide, the proportion of the disabled is higher in rural areas than in urban areas. The disability rate per 100,000 population for the country as a whole works out to 2,215. This is 2,404 in the case of males and 2,017 in the case of females. There are mainly five types of disabilities found in India, i.e. in vision, in movement, mental, in speech and in hearing. Among the five types of disabilities on which data has been collected by the census, disability in vision, which was nearly 48.5 per cent within the group, has emerged as the top category. It is followed by locomotor disability (27.9 per cent), mental disability (10.3 per cent), speech related disability (7.5 per cent), and hearing disability (5.8 per cent). The gender distribution also follows a similar pattern, except that the proportion of disabled females is higher in both vision and hearing categories.

As already mentioned, the present chapter draws from the official secondary data sources. Data for the study comes mainly from the Census of India, and available census data for the year 2001 and 2011 has been used for analysis. The collected data has been classified into various groups and used to make a comparative analysis for the study. The status of disability across social groups as well as their educational status by sex and residence has been analysed. The study traced out the changing structure of disabled persons in India between 2001 and 2011 in terms of proportion to the respective total populations of states. A comparative analysis has been done to understand the literacy status and educational levels across social groups in India. Literacy status by sex and residence across social groups is also part of the analysis.

Disability: a general picture on nature and extent

Understanding disability is indeed a difficult task, given its myriad manifestations and intensity. The World Health Organisation has offered a definition which states that, "disability is any restriction or lack (resulting from an impairment) of ability to perform in a manner or within the range considered normal for a human being". Disability Act, 1995 defines a disabled person as "a person suffering from not less than forty per cent of any disability as certified by a medical authority as disabled". The National Sample Survey Organisation (NSSO) defines disability as "Any restriction or lack of abilities to perform an activity in the manner or within the range considered normal for a human being". It excludes illness/injury of recent origin (morbidity) resulting in temporary loss of ability to see, hear, speak or move.

As per the 2001 Census, the number of disabled persons in India as a whole was 21.9 million (2.13 per cent of the total population), which increased to 26.8 million (2.21% per centof the total population) by 2011 (see Table 9.1). It indicates that the overall disability trend in India is on an upward trajectory and the total disabled persons to the total population of India will be 32.8 million and 40.71 million by 2021 and 2031 respectively. Out of the total disabled persons enumerated in 2011, 16.38 million were in rural India and their number increased to 18.63 million by 2011. As many as 5.51 million persons with disability were in urban India as of the 2001 Census data, which increased to 8.17 million by 2011. Although the share of the disabled has been notably higher in rural India, the increase of disabled persons in urban India was slightly higher (0.24 per cent) than the rural (0.03 per cent). Table 9.1 also provides the gender dimensions of disability. For instance, the share of male disabled persons is higher than females. The number of female disabled persons was 9.30 million in 2001 which increased to 11.82 million in 2011. However, the increase in the number of the disabled has been more for women than men between 2001 and 2011. The increased proportion between 2001 and 2011 for males was 0.04 per cent and in the case of females, it was 0.14 per cent. The percentage of rural

Table 9.1 Key aspects of disabled population of India (2001 and 2011)

	Total		*Rural*		*Urban*		*Male*		*Female*	
	2001	*2011*	*2001*	*2011*	*2001*	*2011*	*2001*	*2011*	*2001*	*2011*
Total Population (in Millions)	1,028	1,210	743	833	286	377	532	623	496	586
Disabled population (in millions)	21.9	26.81	16.38	18.63	5.51	8.17	12.6	14.98	9.3	11.82
% to the total			74.79	69.48	25.21	30.52	57.53	55.87	42.47	44.13
Percentage of disabled population to the total population of India	2.13	2.21	2.2	2.23	1.92	2.16	2.36	2.4	1.87	2.01
Change between 2001 and 2011 (%)		0.08		0.03		0.24		0.04		0.14

Source: Census of India (2001 and 2011). Note: Parenthesis represents the decline/increase of disability in terms of percentage over ten years.

Note: Arrived at by the author through census data

disabled persons to the total population was almost similar for male and female, at 2.20 per cent. However, the number of persons with disability increased more among women in urban areas.

The data on the disabled population in India by state and residence (rural/urban) is presented in Table 9.2. This data shows that the number of disabled persons notably increased in states including Goa, Maharashtra, Manipur,

Table 9.2 Growth rate of disabled population in India by states (2011)

Disabled Population in India by States – 2011				*Total Growth Rate in % (2011)*	*Rural Growth Rate in % (2011)*	*Urban Growth Rate in % (2011)*
State	*Total*	*Rural*	*Urban*			
Jammu & Kashmir	3,61,153	2,73,952	87,201	19.32	19.26	19.53
Himachal Pradesh	1,55,316	1,42,665	12,651	–0.41	–1.44	13.02
Punjab	6,54,063	4,30,313	2,23,750	54.07	44.88	75.48
Chandigarh	14,796	346	14,450	–4.78	–79.11	4.09
Uttarakhand	1,85,272	1,34,097	51,175	–4.88	–13.39	28.13
Haryana	5,46,374	3,57,663	1,88,711	20.07	5.26	63.74
Rajasthan	15,63,694	12,19,186	3,44,508	10.74	9.90	13.85
Uttar Pradesh	41,57,514	31,66,615	9,90,899	20.39	17.03	32.55
Bihar	23,31,009	20,46,351	2,84,658	23.49	20.91	45.86
Sikkim	18,187	15,859	2,328	–10.70	–14.60	29.62
Arunachal Pradesh	26,734	22,308	4,426	–19.75	–18.75	–24.46
Nagaland	29,631	23,549	6,082	11.82	0.12	104.16
Manipur	58,547	40,488	18,059	106.33	93.21	143.35
Mizoram	15,160	8,627	6,533	–5.32	–10.98	3.37
Tripura	64,346	44,942	19,404	9.17	–7.91	91.36
Meghalaya	44,317	37,566	6,751	53.86	65.20	11.35
Assam	4,80,065	4,15,457	64,608	–9.47	–11.25	3.89
West Bengal	20,17,406	13,68,942	6,48,464	9.22	1.08	31.56
Jharkhand	7,69,980	5,97,544	1,72,436	71.73	71.25	73.39
Odisha	12,44,402	10,67,739	1,76,663	21.84	21.65	23.00
Chhattisgarh	6,24,937	4,98,228	1,26,709	48.83	44.42	69.18
Madhya Pradesh	15,51,931	11,05,704	4,46,227	10.18	4.27	28.19
Gujarat	10,92,302	5,70,479	5,21,823	4.48	–20.70	60.02
Maharashtra	29,63,392	16,66,135	12,97,257	88.80	63.29	136.20
Andhra Pradesh	22,66,607	15,12,961	7,53,646	66.05	44.04	139.57
Karnataka	13,24,205	7,92,694	5,31,511	40.78	19.90	90.16
Goa	33,012	14,622	18,390	109.61	70.84	155.77
Kerala	7,61,843	4,14,875	3,46,968	–11.50	–35.95	62.89
Tamil Nadu	11,79,963	6,21,745	5,58,218	–28.16	–34.21	–19.97
India	26,814,994	18,636,358	8,178,636			

Source: Classified by the author using Census data.

Jharkhand and Andhra Pradesh. While the states of Himachal Pradesh, Chandigarh, Uttarakhand, Sikkim, Arunachal Pradesh, Mizoram, Assam, Kerala and Tamil Nadu registered negative or nil growth rate in the disabled population, the rest of the states saw a moderate increase in the number of the disabled population during the decade 2001–2011. The data also shows the rate of increase across rural and urban areas of these states. The states of Himachal Pradesh, Uttarakhand, Sikkim, Arunachal Pradesh, Mizoram, Tripura, Assam, Gujarat, Kerala and Tamil Nadu showed a decrease in the number of disabled population in rural areas between 2001 and 2011. However, the number of disabled persons significantly increased in the rural areas of Manipur (93 per cent), followed by Jharkhand (71 per cent), Goa (70 per cent), Meghalaya (65 per cent), Maharashtra (63 per cent), Punjab (44.8 per cent), Chhattisgarh (44.4 per cent), Andhra Pradesh (44 per cent), Odisha (21.7 per cent), Karnataka (19.9 per cent) and Jammu and Kashmir (19.3 per cent). The number has increased in the urban areas of almost all states except Arunachal Pradesh and Tamil Nadu during the decade.

Intersections of disability, caste and gender

Although studies on disabilities received academic attention, their overarching relations with other marginalities like caste and gender are not adequately investigated in India. We borrow from the concept of multiple marginalisation, which scholars used to study the layers of deprivation of Dalit women (Rege, 2006) and women in distress conditions (Manorama, 2008; Irudayam, Mangubhai, & Lee, 2014) here to refer to the complexities of the problems faced by disabled people who are from the lower rung of the caste/class hierarchy.

Pal (2010), using NSSO 50th round data on *Disabled Persons in India*, estimated that nearly one-third of the disabled population in India was below 15 years and the share of disabled in the productive age group was more among SCs and STs as compared to the upper castes. He has also revealed that disabilities caused due to avoidable conditions such as anaemia, pneumonia and malnutrition were more for SCs and STs as compared to upper castes. Although Pal's study did not find significant differences in the prevalence of disability across social groups, it pointed out the differential impacts of disability for Dalits since it adds another layer of deprivation to their already disadvantaged conditions.

The data that we extracted from the census, however, showed the interlocking characteristics of caste and disability. Table 9.3 presents the data on disabled persons across social groups and sectors for the years 2001 and 2011. The data clearly shows the rural–urban and caste divide of the disabled population in India. It shows that nearly 90 per cent of disabled persons from ST and 77 per cent from SC population lived in rural areas in 2011. As is evident from the table, the share of disabled persons decreased

Table 9.3 Disabled persons in India across social groups by residence (2001 and 2011)

Social Groups	*Total Persons*		*% to the Total*		*% of Rural*		*Change between 2001–2011*	*% of Urban*		*Change between 2001–2011*
	2001	*2011*	*2001*	*2011*	*2001*	*2011*		*2001*	*2011*	
SCs	37,11,110	49,27,433	16.94	18.38	81.81	77.09	–4.72	18.18	22.9	4.72
STs	16,18,166	21,40,763	7.39	7.98	92.74	90.02	–2.72	7.25	9.97	2.72
Others	1,65,77,493	1,97,46,798	75.67	73.64	71.49	65.37	–6.12	28.51	34.62	6.11
Total	2,68,14,994	2,19,06,769	2.1	1.8	74.8	69.49	–5.31	25.19	30.5	6.11

Source: Census data, extracted.

Table 9.4 Disabled persons in India across social groups by sex (2001 and 2011)

Social Groups	*Total Persons*		*% to the Total*		*% of Male Persons*		*Change between 2001–2011*	*% of Female Persons*		*Change between 2001–2011*
	2001	*2011*	*2001*	*2011*	*2001*	*2011*		*2001*	*2011*	
SCs	37,11,110	49,27,433	16.94	18.38	57.33	56.22	–1.11	42.66	43.77	1.11
STs	16,18,166	21,40,763	7.39	7.98	55.85	53.46	–2.39	44.14	46.53	2.39
Others	1,65,77,493	1,97,46,798	75.67	73.64	57.75	56.07	–1.68	42.24	43.92	1.68
Total	2,68,14,994	2,19,06,769	2.1	1.8	57.74	55.89	–1.65	42.45	44.1	1.68

Source: Census data, extracted.

among all social groups in rural India and increased in urban India. What is important pertaining to the social group difference is that while the share of disabled persons as a whole has come down in non-SC/ST groups between 2001 and 2011, it increased for SCs and STs. This is accompanied by two trends. First, although there was a general decline in the number of the disabled in all social groups in rural areas, it was less pronounced for SCs and STs as compared to non-SC/STs. Second, like in all social groups, there was an increase in the number of the disabled among SCs and STs as well. This could also be interpreted either as part of the general urbanisation process and as the increasing movement of persons with disabilities from these groups to urban areas. However, it requires more investigation to unearth the increase of the disabled in the urban areas in general and among SCs and STs in particular.

Gender is another important intersecting category adding to the complexity of disability. Table 9.4 provides data of disabled persons across social groups and gender as well as the changes over a decade. As already discussed, while the share of the disabled decreased in non-SC/ST groups, it increased among SCs and STs. Table 9.4 helps to explain lucidly the gender differentials of this trend. While the number of men with disability decreased among SCs and STs, the number of women increased, which in fact explained the general increase in the share of the disabled in these groups. This gendered trend was true for all social groups; however, it was more pronounced for STs.

Disability and educational mobility: caste and gender differentials

Education is an important component of human capital formation. The affirmative action policies of the state (Thorat & Senapati, 2007) and the various initiatives of lower caste groups from below have significantly contributed to the educational mobility of various lower caste groups in India. Pal (2010) examined the educational status of disabled people across social groups, which is presented in Table 9.5. We reproduce this data here to start our discussion on the educational mobility and advancement of disabled persons in general and the disabled in the lower caste groups in particular, with their current educational status. As is clear from the table, among the physically disabled persons, the share of non-literates was the highest for STs followed by SCs. Besides, the share of non-literate females was notably higher than males in all the social groups and interestingly the male–female differential was the highest among OBCs, followed by upper caste groups, SCs and STs. Another interesting finding is that although illiteracy was the highest among women in all social groups, the share of male illiterate members in ST groups was slightly more than the illiterate women from upper caste groups. Also, the share of disabled SCs and STs in the groups of the

Table 9.5 Educational status of persons with disability by social group and gender

General Education	*SC*		*ST*		*OBC*		*Others*		*Total*	
	M	*F*	*M*	*F*	*M*	*F*	*M*	*F*	*M*	*F*
Not literate	54.6	76.3	60.1	79.8	47.4	72.2	35.8	59.9	46.2	70.1
Literate without schooling	1.4	.9	1.9	1.2	1.5	.8	1.9	1.9	1.7	1.1
Literate but below primary	13.4	8.3	12.6	8.2	14.4	9.3	12.3	10.3	13.5	9.2
Primary	13.6	6.8	11.3	6.1	13.8	8.0	14.3	10.6	13.8	8.4
Middle	10.5	5.2	8.3	3.1	12.5	5.7	16.8	8.4	13.1	6.2
Secondary	3.5	1.2	2.9	.8	5.5	2.2	8.9	4.8	5.9	2.6
Higher secondary	1.7	.8	1.7	.5	2.7	1.1	5.0	2.0	3.1	1.3
Diploma/certificate course	.2	.1	.3	.0	.4	.2	.7	.3	.4	.2
Graduate	.8	.3	.7	.3	1.3	.5	3.3	1.4	1.8	.7
Post graduate and above	.2	.1	.1	.1	.4	.2	1.0	.4	.5	.2

Source: Pal (2010).

educated, including all categories from primary to graduation, was lower than that of non-SC/STs with similar gender differentials.

Over the years, there has been an improvement in the educational status of the disabled population in the country. This is true for both males and females. For instance, the overall literacy level of the disabled in India increased over time, i.e. from 50.7 per cent (2001) to 55.4 per cent (2011). Also, persons with a graduation degree and above among males were 6.6 per cent in 2001 which increased to nearly 9 per cent by 2011. Similar improvement was visible for females as well (4.6 per cent in 2001 and 7.7 per cent in 2011). In the following section, we examine the educational mobility of the disabled population across social groups with a focus on rural–urban and gender differentials.

A comparison in terms of literacy status and educational level among literates by gender between 2001 and 2011 is presented in Table 9.6. Out of 26.8 million disabled people in India, 14.6 million are literates (54.5 per cent) and the remaining 12.2 million (45.5 per cent) are illiterates. The percentage of literates in the disabled population was 49.3 per cent in 2001 which went up to 54.5 per cent in 2011 with an increase of 5.2 percentage points. The male–female differentials in educational disparities, although declining between 2001 and 2011, continued to be significant. The increase in the share of literate women in the disabled population is indeed

Table 9.6 Percentage distribution of disabled population by literacy status and educational level in India (2001 and 2011)

Literacy status and educational level among disabled population	*2001*			*2011*			*Progress between 2001 & 2011*		
	Total	*Male*	*Female*	*Total*	*Male*	*Female*	*Total*	*Male*	*Female*
Illiterate	50.7	41.9	62.7	45.5	37.6	55.4	–5.2	–4.3	–7.3
Literate	49.3	58.1	37.3	54.5	62.4	44.6	5.2	4.3	7.3
Percentage to literates							0	0	0
Literate but below primary	26.5	25	29.9	19.4	18.5	21.5	–7.1	–6.5	–8.4
Primary but below middle	26.4	25.6	28.1	24.3	23.5	25.8	–2.1	–2.1	–2.3
Middle but below matric/secondary	16.4	16.6	14.8	16.7	17.3	15.8	0.3	0.7	1
Matric/secondary but below graduate	20	21.5	16.9	23.6	24.9	21.2	3.6	3.4	4.3
Graduate and above	6	6.6	4.6	8.5	9	7.7	2.5	2.4	3.1

Source: Registrar General of India, Ministry of Home Affairs, Govt. of India. July 2016.

a welcome outcome. The data also shows a positive picture of educational attainment among the disabled population reflected in terms of their participation in secondary and above level. Most interestingly, although males outnumbered females in all levels of educational attainments, the increase was visible more for females with secondary education and above, which in a way would help reduce the gender differentials in educational attainments.

If we take the level of literacy by social groups, it is evident that the STs and SCs are the more vulnerable groups within the disabled. As per the latest census, among the disabled, the share of ST and SC literates to the total literates was around 6 per cent and 16.18 per cent respectively (see Table 9.7). There is a wide gap of literates across social groups in India between 2001 and 2011. The gap is almost 12 percentage points between the general category and SCs, and around 7 percentage points between SCs and STs. This means attained literacy among STs is lower when compared to other groups. Here the significant fact to be noted is that among the total disabled persons, a majority of the disabled (almost 81 per cent) literates belong to the non-SC/ST group. Altogether, SC/ST persons constitute only 20 per cent of the literates among the total disabled population in India. The situation has improved marginally for SCs between 2001 and 2011, but the improvement was not pronounced for STs.

The literacy status of disabled persons across social groups in rural and urban areas is presented in Table 9.7. The data shows that the literacy status among disabled SC male persons in 2001 was 50 per cent which increased to 56.26 per cent in 2011. Among STs, it is 43.31 per cent in 2001 which increased to 49.53 per cent by the year 2011. The increase in the proportion of literates between 2001 and 2011 among SCs and STs is almost similar, with 6.16 per cent for SCs and 6.22 per cent among STs respectively. But it should be borne in mind that the literacy status of ST males is lower than that of SC males. The difference is around 6 per cent.

Table 9.7 Status of literacy among disabled persons in India across social groups by residence (2001 and 2011)

Social Groups	*Total Literate Persons*		*% to the Total*		*% of Rural Persons*		*Change between 2001–2011*	*% of Urban Persons*		*Change between 2001–2011*
	2001	*2011*	*2001*	*2011*	*2001*	*2011*		*2001*	*2011*	
SCs	15,17,724	23,65,085	14.05	16.18	38.15	44.57	6.42	53.24	59.51	6.27
STs	54,812	8,69,729	5.07	5.95	32.32	38.62	6.3	53.35	58.66	5.31
Others	87,35,696	1,13,83,539	80.88	77.87	47.53	51.68	4.15	65.64	68.91	3.27
Total	1,08,01,232	1,46,18,353	49.3*	54.51*	67.36	62.32	–5.04	32.63	37.67	5.04

Source: Census data.

Note: Literate persons to the total disabled persons of India.

The attainment of literacy among disabled persons with respect to females across social groups has confirmed that the overall increase of female literacy is lower than that of males. This can be seen in Table 9.8. The observed SC female literacy in the year 2001 was 28 per cent which increased to 37.38 per cent in 2011 with an increase of nearly 9 percentage points. Similarly, the literacy rate among disabled females from STs in 2001 was 21.87 per cent which went up to 30.39 per cent by 2011 with an increase of nearly 9 percentage points. However, the gender differentials appear to have widened when we examine educational and literacy status across social groups. This is reflected at three levels. First, as already highlighted, there was a notable difference between males and females within the social groups. As is evident from Table 9.8, gender differentials in literacy status among the disabled was the highest among STs, followed by SC and non-SC/STs. Second, among females, the share of illiteracy was the highest for STs, followed by SCs. Third, among disabled people, the gender differential in literacy status was the highest between ST women and non-SC/ST men followed by SC women and non-SC/ST men.

Disability and occupation

Pal (2010), based on the NSSO 50th round data on disabled persons in India, estimated the occupational status of people with disability across social groups, which is presented in Table 9.9. He showed that a majority of the people with disability are engaged in unrecognised and economically unrewarding works and only 25 per cent are engaged in some form of productive occupation. Their share in salaried occupation within the group was as low as 3 per cent. Pal also highlighted the gender and social group differentials in the occupational status of people with disability. He observed that the share of women in economically rewarding activities was "substantially lower" than men. Similarly, the share of persons in employment was lower for SCs and STs as compared to "other" social groups.

Census 2011 provides the latest data on the employment status of persons with disability (see Table 9.10). Interestingly, the proportion of workers among the disabled is found to be the highest among STs (42 per cent) followed by SCs (38 per cent) and others (35 per cent). It should also be mentioned that the census data on workers and non-workers comprise all who work for a living, irrespective of their ability to work. Hence, the higher share of workers from STs and SCs also implies their lower levels of support systems. In line with the general trend, the share of women workers is less than men in all social groups. However, the share of women workers among the disabled population is found to be the highest for STs (35 per cent) followed by SCs (25 per cent) and then others (21 per cent).

Table 9.8 Literacy status across gender and social groups of disabled population in India (2001 and 2011)

Literacy status	*SC*											
	2001						*2011*					
	Male	%	*Female*	%	*Total*	%	*Male*	%	*Female*	%	*Total*	%
Literate	1066083	50.10	451641	28.5	1517724	40.90	1558834	56.26	806251	37.4	2365085	48
Illiterate	1061794	49.9	1131592	71.5	2193386	59.10	1211758	43.73	1350590	62.6	2562348	52
	ST											
	Male	%	*Female*	%	*Total*	%	*Male*	%	*Female*	%	*Total*	%
Literate	391554	43.3	156258	21.9	547812	33.85	566970	49.53	302759	30.4	869729	40.63
Illiterate	512345	56.7	558009	78.1	1070354	66.15	577657	50.46	693377	69.6	1271034	59.37
	Others											
	Male	%	*Female*	%	*Total*	%	*Male*	%	*Female*	%	*Total*	%
Literate	5819118	60.8	2916578	41.6	8735696	52.70	7222549	65.22	4160990	48	11383539	57.65
Illiterate	7537488	39.2	304309	58.4	7841797	47.30	3850825	34.78	4512434	52	8363259	42.35

Source: Census of India, Report on Disability (2001 and 2011).

Table 9.9 Work/activity status of persons with disabilities across social groups

Work/Activity Status	*Dalits*	*ST*	*OBC*	*Others*	*Total*
Work in household enterprise (self-employed): Own account worker	7.2	8.4	10.2	11.2	9.7
Employer	0.1	0.1	0.3	0.5	0.3
Work as helper in household enterprise (unpaid family worker)	2.5	7.0	4.6	4.6	4.3
Work as salaried/ wage employee	2.4	1.8	2.7	3.9	2.9
Work as casual wage labour: in public work	0.2	0.1	0.1	0.0	0.1
In other types of work	12.2	11.0	6.6	4.7	7.6
Do not work but seeking and/or available for work	0.9	0.3	0.6	0.8	0.7
Attend educational institution	14.8	12.5	14.1	13.5	13.9
Attend domestic duties only	9.0	8.1	10.2	11.0	10.0
Attend domestic duties, engaged in free collection	2.7	3.2	2.5	2.2	2.5
Rentiers, pensioners, rémittence récipients etc	1.4	1.4	1.8	2.5	1.9
Not able to work due to disability	32.1	28.6	31.0	30.6	31.0
Beggars, prostitutes etc	0.8	0.8	0.5	0.5	0.6
Others	13.7	16.7	14.7	13.8	14.4

Source: Pal (2010).

Table 9.10 Distribution of disabled workers and non-workers in India across social groups, 2011

Employment status	*Male*	%	*Female*	%	*Total*	%
	SC					
Workers	13,39,451	48.35	5,45,473	25.29	18,84,924	38.25
Non-Workers	14,31,141	51.65	16,11,368	74.71	30,42,509	61.75
Total	27,70,592	100	21,56,841	100	49,27,433	100
	ST					
Workers	5,55,180	48.50	3,46,241	34.76	9,01,421	42.11
Non-Workers	5,89,447	51.50	6,49,895	65.24	12,39,342	57.89
Total	11,44,627	100	9,96,136	100	21,40,763	100
	Others					
Workers	51,78,194	46.76	17,79,847	20.52	69,58,041	35.24
Non-Workers	58,95,180	53.24	68,93,577	79.48	1,27,88,757	64.76
Total	1,10,73,374	100	86,73,424	100	1,97,46,798	100

Source: Census of India (2011).

Concluding remarks

The data presented in this chapter unravels the multiple deprivations of disabled persons in India with regard to their educational attainments and upward mobility. The extent of disability is the leading question that the chapter attempted to understand. Our analysis showed that there has been a considerable increase in the number of disabled persons in India over the years. States like Uttar Pradesh, Maharashtra, Andhra Pradesh, Bihar, West Bengal, Madhya Pradesh, and Tamil Nadu have a higher incidence of disability and they account for a large proportion of the disabled population in the country. The percentage increase among the urban population to the total was slightly higher than the rural. It is to be noted that in absolute terms, male disability has declined over time, whereas the decline has been slower among females. One major thing that needs to be noted here is that disability among rural persons has reduced over time as compared to their urban counterparts. We could also identify gradual improvements in the capabilities, mostly in terms of education among the disabled persons in general. For instance, as already discussed in the chapter, during the last two decades, not only has the overall educational status of the disabled population improved, but the percentage of literates among both the male and the female disabled population has also increased; and their status in higher educational levels has also improved across all social groups. The decade 2001–2011 also witnessed an increase in educational attainments among the disabled females compared to their male counterparts.

Across social groups, increase in the share of the disabled was found only among STs between 2001 and 2011. It should be mentioned that the share of non-SC/STs in the disabled population has declined marginally during this period. Similarly, although the overall educational attainments of the disabled population have improved over a period of time, it was not significant for SCs and STs.

The Union Ministry of Social Justice and Empowerment is the nodal agency that promoted services for people with disabilities through its various schemes such as the Regional Rehabilitation Training Centre (RRTC) till 1985. The National Information Centre on Disability & Rehabilitation (NICDR) came up in 1987, and subsequently, entities like the National Council for Handicapped Welfare, National Handicapped Finance & Development Corporation, etc. emerged. The main aim of the ministry was to promote services for disabled persons through government and non-government organisations, so that the disabled are encouraged to become functionally independent and productive members of the nation through opportunities for education, vocational training, medical rehabilitation and socio-economic rehabilitation. This chapter suggests that the above schemes and plans should be further fine-tuned to work towards the betterment of the disabled population in their social, educational and economic empowerment in general, and social security in particular.

References

Census of India. (2011). *Provisional population totals*. New Delhi: Office of the Registrar General & Census Commissioner, Government of India.

Chaudhuri, L. (2006, May). *Disability in India: Issues and concerns, ess conference paper*. Retrieved May 21, 2018, from www.apangutkarsh.com/pdf/disabilityinindia.pdf

Groce, N., Kett, M., Lang, R., & Trani, J. F. (2011). Disability and poverty: The need for a more nuanced understanding of implications for development policy and practice. *Third World Quarterly*, *32*(8), 1493–1513.

Irudayam, A., Mangubhai, J., & Lee, J. (2014). *Dalit women speak out: Caste, class, gender violence in India*. New Delhi: Zubaan.

Jenkins, R. (1991). Disability and social stratification. *The British Journal of Sociology*, *42*(4), 557–580.

Manorama, R. (2008). Dalit women: The downtrodden among the downtrodden. In M. E. John (Ed.), *Women's studies in India: A reader* (pp. 445–452). New Delhi: Penguin Books India.

Pal, G. C. (2010). *Dalits with disabilities: The neglected dimension of social exclusion* (Draft Report: Status of persons with disability in India: Comparison across social groups Working Paper). New Delhi: Indian Institute of Dalits Studies. Retrieved January 12, 2018, from www.dalitstudies.org.in/uploads/publication/1473143384.pdf

Rege, S. (2006). *Writing caste, writing gender*. New Delhi: Zubaan Books.

Registrar General of India. (2016). *Data on disabled population 2001 and 2011*. New Delhi: Ministry of Home Affairs, Government of India.

Thorat, S., & Senapati, C. (2007). *Reservation in employment, education, and legislature: Status and emerging issues*. New Delhi: Indian Institute of Dalit Studies.

10

ECONOMIC, OCCUPATIONAL AND LIVELIHOOD CHANGES OF SCHEDULED TRIBES OF THE NORTH EASTERN REGION[1]

Reimeingam Marchang

Introduction

Scheduled Tribes (STs) predominantly live in the hills of North Eastern region (NER),[2] depending on agriculture as their main source of livelihood and income. They are considered to be historically socially disadvantaged and economically underdeveloped people. They comprise about 27 per cent of the entire NER's population.[3] Arunachal Pradesh (Ar.P), Meghalaya, Mizoram and Nagaland among the NE states are ST population-dominated states with over two-thirds of each state's total population. A transition of their livelihood system from agricultural to non-agricultural pursuits is evident. They share several disadvantages, including geographical isolation, underdevelopment, economic deprivation, illiteracy, impoverishment, indebtedness and lower access to assets and public services (Hanumantha & Grover, 1979). They have traditionally lived mainly in forests, hills and undulating inaccessible terrain in plateau areas that have rich natural resources (Ministry of Tribal Affairs, 2013). They have been at the "lower end in all indicators of living conditions and household assets" (Bhagat, 2013, p. 64) mainly due to the government's adoption of unbalanced development programmes.

Nevertheless, generalisation of STs across India would be erroneous due to differences in development outcomes such as education, health and income across the Indian states. Chaube (1999) cautioned that STs of India are heterogeneous groups. Hence, the common definition of "tribe" or "tribal" does not exist because of the difference in the social, economic and administrative structures, customary norms, value system, geographical isolation and nature of underdevelopment. "Tribal" is a way of living or living system (Sikidar, 1990). Tribal societies are organised according to cultural principles and tribes are distinguished from one another by kinship and lineage but not by occupation (Corbridge, 1988). On the contrary, the

Indian occupational system is deeply rooted within the Hindu caste system (Horan, 1974). ST people do not have a traditional social hierarchy similar to Hindu social or caste hierarchy (Sundaram & Tendulkar, 2003; Bhagat, 2013). However, they are a marginalised class (Roy, 1989) and are relatively deprived with respect to non-tribal people in many aspects (Srivastava, 2008). Their social and economic underdevelopment is due to habitation in geographically isolated areas in rough terrains and the practice of shifting or jhum cultivation for their livelihood (Sundaram & Tendulkar, 2003). Shifting cultivation was the major means for their livelihood (Christoph, 1982; Kumar & Ramakrishnan, 1990; Nongbri, 1999; Sundaram & Tendulkar, 2003; Sengupta, 2013). However, in recent times, their means of livelihood has increasingly oriented towards non-agricultural activities (Marchang, 2016). With this backdrop, this chapter[4] examines the gradually changing economic, occupational and livelihood conditions orienting STs of the region towards non-agricultural activities.

Data sources and methodology

This study attempts to produce and include both qualitative and quantitative results and analyses them descriptively while emphasising on the means of livelihood of the STs of NER. It is based on recent secondary data from NSSO, Census and CSO, and primary field survey data. Based on this secondary data, the patterns and trends of the livelihood system focusing on economic groups, economic structure and the employment structure of STs of NER are examined. Additionally, primary data, i.e. detailed socio-economic profiles including livelihood means, was collected by randomly drawing 170 households from the randomly drawn state of Manipur (among NE states) through a field survey during March to August 2014 to exhibit the livelihood status of STs of Manipur.[5] Manipur has the characteristics of NER because of its population structure (which includes ST, SC, OBC and others) and economic condition. Samples were drawn from all the five hill districts namely Chandel, Churachanpur, Senapati, Tamenglong and Ukhrul, where different Scheduled Tribes population predominated, and three valley districts, Imphal East, Imphal West and Thoubal, where the ST population are sparingly scattered. Samples were randomly drawn from different proximities to the urban centres of these districts. To canvass the information, a personal interview method was adopted using a semi-structured household questionnaire. The results of secondary data are strengthened and validated by the results of primary data to a certain extent.

Economic groups or household types

NSSO data, which we have used in the present chapter, classifies households in rural and urban areas separately depending on the sources of income.

Household type is an economic grouping (Sundaram & Tendulkar, 2003). According to NSSO (2015), household type classification is based on the major sources of the household's income or livelihood during the one year preceding the date of survey. Only the net income from economic activities was considered in determining the household's income. In rural areas, the households were classified into self-employed in agriculture, self-employed in non-agriculture, rural labour in agriculture, others (non-agricultural) rural labour and the others category until 2009–2010. Till 2009–2010, the others category rural households include households with regular wage/salary. In 2011–2012, rural households whose major source of income was from regular wage/salary was separated from the others category. As per Sundaram and Tendulkar (2003) land is the major capital and source of livelihood for the self-employed in an agricultural household. The major source of livelihood for the self-employed in a non-agricultural household is the deployment of non-agricultural physical or human capital assets in the production process. The means of livelihood for agricultural rural labour households depend on their endowments of abundant manual labour. In the case of the other (than agricultural) category of rural labour households, the non-agricultural manufacturing or service activities on non-contractual casual basis constituted the main source of livelihood. The others category households include both households whose major source of income arises mostly from contractual employment with regular wages and salaries, and who earn their living from non-labour assets without direct participation in gainful economic activity. Non-participatory earnings from non-labour assets include current returns from the ownership of immovable assets from land or real estate, returns from past financial investments, and receipts from public or private transfers (including pension and remittances).

As presented in Table 10.1, in rural areas, a majority of the ST households were classified as self-employed in agriculture and non-agriculture for all the NE states excepting Tripura. In Tripura, 45 per cent of households were self-employed in agricultural and non-agricultural households in 2011–2012. The shares of self-employed households were much higher for all the NE states, excepting the state of Tripura, when compared to the national level. This is because a large share of the country's rural households, unlike in NE, were rural labour households. The share of rural labour households was smaller in all the NE states, excepting Tripura, especially in the ST dominated states like Arunachal Pradesh, Mizoram and Nagaland. The share of rural labour households was low in Manipur – the ST minority state. Interestingly, the share of the others (including wage/salary) category of households was greater for NE states, excepting Assam, when compared with the same for India in 1999–2000. A similar situation was present in 2011–2012.

Expectedly, the size of agricultural households (combining both self-employed in agriculture and agricultural labour) has declined in all the

Table 10.1 Distribution (%) of household type of ST in rural/urban areas of NE states/India

State/country	*Year*	*Rural*						*Urban*			
		Self Employed		*Rural Labour*		*Regular wage/ salary*	*Other including wage/ salary*	*Self Employed*	*Wage/ Salaried*	*Casual Labour*	*Others*
		Agri	*Non Agri*	*AL*	*OL*						
Ar.P	1999–00	52.4	8.6	7.6	2.6	–	28.7	11.5	53.4	3.6	31.5
	2011–12	72.2	5.1	0.2	0.9	13.6	21.5	22.2	60.9	7.4	9.5
Assam	1999–00	63.4	7.1	11.7	9.7	–	8.1	20.0	40.7	5.2	34.1
	2011–12	52.6	21.9	10.3	5.2	4.4	9.9	41.2	43.7	0.7	14.4
Manipur	1999–00	77.8	3.7	4.4	1.9	–	12.3	29.3	20.7	7.5	42.5
	2011–12	53.2	19.4	2.7	1.3	21.2	23.4	54.3	23.3	3.3	19.2
Meghalaya	1999–00	69.3	5.6	10.2	4.1	–	10.3	17.3	48.7	13.5	20.0
	2011–12	43.9	20.7	11.2	5.2	15.3	19.0	25.2	45.5	13.2	16.1
Mizoram	1999–00	61.6	5.5	5.9	2.9	–	24.2	36.8	42.5	12.2	8.2
	2011–12	69.8	10.8	0.0	2.7	15.3	16.7	34.8	52.0	7.2	5.9
Nagaland	1999–00	59.4	1.9	3.7	12.0	–	32.9	7.0	67.1	0.4	25.5
	2011–12	57.8	12.3	0.0	0.4	21.8	29.5	22.3	64.3	1.9	11.5
Sikkim	1999–00	49.4	1.9	3.7	12.0	–	32.9	18.4	58.0	6.3	17.4
	2011–12	56.9	8.5	0.0	7.4	23.6	27.1	39.4	35.7	0.0	24.9
Tripura	1999–00	23.6	8.3	28.6	15.6	–	23.9	4.7	80.3	0.0	15.0
	2011–12	29.6	15.7	5.3	35.1	6.7	13.8	12.2	39.7	0.9	47.2
India	1999–00	36.2	5.2	39.7	8.9	–	10.1	21.6	38.0	25.6	14.7
	2011–12	41.4	8.1	24.5	13.9	6.3	12.2	19.5	46.5	18.0	16.0

Source: NSSO (2001, 2015).

Notes: AL – agriculture labour, OL – other labour. Agri – agriculture. Figures given per 1000 in NSSO Reports are converted into percentage. -- Not available. Rural household category such as regular wage/salary was added only in 2011–2012.

NE states following the national trend, except Ar.P, Mizoram and Sikkim where it has marginally grown. The share of rural labour, i.e. agricultural rural labour and other rural labour households have declined for all the NE states, except Meghalaya, following the national trend during 1999–2000 to 2011–2012. The share of the others (including wage/salary) category of households has declined for Arunachal Pradesh, Mizoram, Nagaland, Sikkim and Tripura against the increasing trend for the country as well as Assam, Manipur and Meghalaya during the same period. A significant share of the NE rural households were regular wage/salary households ranging from 4 per cent in Assam to about 24 per cent in Sikkim against the 6 per cent of India in 2011–2012. Wage and salary households contributed most of the other category households.

This indicates a change in economic pursuits from agriculture to non-agricultural activities for most of the NE states.[6] The unavailability of non-agricultural employment or increase in cash crop farming may explain this situation for those states that have experienced an increase in agricultural households. On the contrary, increase in availability of non-agricultural employment due to economic development explains the increase of non-agricultural households.

In urban areas, households were classified as self-employed, wage or salaried, casual labour and others. Sundaram and Tendulkar (2003) remarked that the self-employed households, including agricultural, are a heterogeneous group that ranges from low income unskilled low productivity trading and personal services with meagre physical or human capital to high income professionals earning their incomes from high skills and education. Further, the others category includes households whose major source of income is derived from non-participatory earnings from non-labour assets.

The share of self-employed households in urban areas was much lower than in rural areas. For STs, it is also the regular wage/salaried households which form the largest share of urban households for all the NE states, excepting particularly Manipur, and the country in all the years (Table 10.1). The share of self-employed households was very significant in the state of Manipur. For example, in 2011–2012, about 54 per cent of the urban households in Manipur were self-employed households, signalling a low generation of salaried employment in the state. Strikingly, in 2011–2012, almost all NE states have a higher share of self-employed households (including agricultural households) in comparison with India which partially means that NE states have a higher dependency on agriculture in urban areas than in the country as a whole. Manipur ranked the least in wage/salaried households with 23 per cent against the top-ranked state Nagaland with about 64 per cent, in 2011–2012.

During 1999–2000 to 2011–2012 the self-employed households have also increased in all the NE states, except Mizoram, following the national trend. Strikingly, casual labour households continue to be lower in all the

NE states than the country's level. Meghalaya has a relatively higher share of casual labour urban households when compared with the other NE state counterparts. Mizoram has the least share of the other category of urban households in comparison with the rest of the NE states and the country's level in all the years. During 1999–2000 to 2011–2012, the shares of casual labour households have declined for all the NE states, except Arunachal Pradesh and Nagaland, as in India. The regular wage/salaried households have increased only in Arunachal Pradesh, Assam, Manipur and Mizoram among the NE states following the national trend. Moreover, the others category of urban households have increased only in Sikkim and Tripura among the NE states following the national trend.

Economic structure: changes and continuities

The economy of NER is "underdeveloped agrarian societies with very weak industrial sectors and inflated service sectors" (Sachdeva, 2000, p. 13). Its "slow pace of urbanisation and extremely lethargic and distorted process of industrialisation have failed to generate alternative means of livelihood for the surplus manpower of the agricultural sector" (Hussain, 2004, p. 4516). Nevertheless, its economy,[7] also entire NE states irrespective of ST or non-ST population dominated states, has gradually changed from a predominant agrarian economy towards an industrial and service economy. The changing economic structure coincides with its employment structure converging towards non-agricultural activities. In 2011–2012, the income contribution from agriculture and allied, industry and services was 23, 23 and 54 per cent respectively for NER (Table 10.2). Marchang (2016) highlighted that income from construction is the major source of income for the industry sector; while income from transport, storage and communication, trade, hotels

Table 10.2 Share (%) of income (real NSDP at factor cost) in NER

States/ Region	*Agri and Allied*		*Industry*		*Services*	
	2000–01	*2011–12*	*2000–01*	*2011–12*	*2000–01*	*2011–12*
Ar. P	35.7	30.1	18.2	30.3	46.1	39.7
Assam	35.9	22.9	16.1	19.3	47.9	57.9
Manipur	31.2	26.3	19.9	33.3	48.8	40.4
Meghalaya	23.8	16.6	22.7	28.8	53.6	54.6
Mizoram	21.1	20.1	14.8	19.2	64.1	60.7
Nagaland	32.9	29.1	12.6	15.8	54.5	55.1
Sikkim	22.7	8.7	22.1	58.1	55.2	33.2
Tripura	25.7	23.9	20.9	24.2	53.5	51.9
NER	33.1	22.8	17.1	22.9	49.8	54.3

Source: Author's calculation based on CSO (2013) and Marchang (2016).

and restaurants, public administration and other services formed the main source of the service sector income. Importantly, income from the entire sub-service sector contribution to the region's income has increased, except for real estate, etc., over the years.

This pattern and trend could be because of various associating factors. The decline of agricultural income is attributed to the decline in the per capita availability of land, especially agricultural land and agricultural employment apart from low production and productivity due to low use of improved seeds and other improved inputs and modern agricultural technology. This results in a shortage of food grains in the region that forces it to import it from the rest of the country to meet the requirement. In order to meet this deficit, the region imports food grains from the rest of India (Hussain, 2004; Husain, 2006). It can be contextualised with the practice of shifting cultivation in the hills with low productivity, especially subsistence production, and the slow progress of modern methods of agriculture in the valleys of Assam and Manipur that are non-ST dominated states.

Marchang (2016) shows that increase in density of workers in agriculture exhibits surplus labour. It lowers productivity and creates disguised unemployment and underemployment. It is expected that the surplus rural labourers migrate towards industrial or urbanised areas to take up non-agricultural employment. However, all the surplus labour from the agriculture sector is not absorbed in the industrial sector as the pace of industrialisation or non-farm employment growth is slow. Inadequate investment in the growth of infrastructure is held responsible for the sluggish growth of industries in the region.[8] Marchang (2016) asserts that income from the service sector increases with an increase in the participation of private players in the market and improvement in productivity due to advances in technology. Moreover, income increases despite the downsizing of the organised sector employment in the government machinery in the post-economic reform period which witnessed privatisation of the public sector and downsizing of the organised sector employment. Interestingly, with the change in economic structure, the real per capita income has also increased from Rs 1,647 in 1990–1991 to Rs 13,146 in 2000–2001 and further rose to Rs 27,562 in 2011–2012 in NER as against India's Rs 2,213 in 1990–1991, Rs 16,688 in 2000–2001 and Rs 38,037 in 2011–2012 (Marchang, 2016). It remains lower for the region than the country's average due to underdevelopment, underdeveloped industrialisation, low productivity and high population growth among other reasons despite the higher educational growth in the region. Goswami (1984) suggested that a persistent rise in per capita income in real terms emanating from increased domestic factor productivity without accentuating economic disparities will bring tribal development in NER. This must be achieved while minimising the adverse effects on future resource availability and ecology and without jeopardising ethnic identity. Lopsided progress in one sphere is often at the expense of the progress of others that is disastrous, particularly for the STs. Therefore, development of STs is envisaged as foremost and crucial for the region's overall development. Additionally, acceleration of income and economic growth depends on investment and level of education (Vaizey, 1967; Stevens & Weale, 2003).

Changes in occupational structure and livelihood

Occupational changes or mobility usually examine the occupational changes from fathers to their children, particularly sons, considering the occupational prestige (Horan, 1974). It is typically an inter-generational occupational change. This section examines the occupational changes over a period of time which may be inferred as inter-generational occupational changes. In North East India, the indigenous people's livelihood means is agriculture and most of them derive their income partly from cultivation and partly from gathering forest products and engaging in other activities (Roy, 1989). Land is an important and indispensable resource for the livelihood of indigenous people. As such, land and forest are fundamental factors for the agrarian economy for STs of NER (Nongbri, 1999). Based on it, shifting cultivation has been practised prominently as a way of life within the tribal communities and hill people from time immemorial (Peale, 1874; Seavoy, 1973; Gupta, 2000); however, all the tribes did not practise it (Corbridge, 1988). The practice occupies a distinct place in the tribal economy and constitutes a vital part of their livelihood means, life-style and socio-economic set-up and is deeply rooted in indigenous ethnic culture (MEF & GBPIHED, 2009; Roy, Xavier, & William, 2012). It is the primary means of livelihood for hill people of the NE states (Thangchungnunga, 1987; Reddy, 1991; Shimray, 2004). Marchang (2016) found that among the ST agriculturists, shifting cultivation is widely practised; but expectedly, the estimated share of shifting cultivation families has declined in recent times. It continues to be more prominent in ST dominated states. However, the area under shifting cultivation is half a hectare per family which is very small as the cultivation is labour intensive. The average area per shifting cultivation family under shifting cultivation was the highest among the ST dominated states like Arunachal Pradesh, Mizoram and Nagaland. Sikkim is the only NE state which does not practise shifting cultivation.

The practice of shifting cultivation would be difficult to eradicate unless alternative means of livelihood are available. In Tripura, alternative means of livelihood such as rubber plantation, horticulture, animal husbandry, dairy and poultry farming were initiated for shifting cultivators (Kumar, 1987). Many such schemes and programmes have been implemented in the past to control shifting cultivation (Maithani, 1991); however, the practice remains widespread, basically due to the hilly topographical conditions (Saikia, 1991). The pressure on shifting agricultural land has sharply increased as cultivation areas are reduced with a simultaneous increase of cultivators.[9]

This occurs despite the fact that the land is owned and controlled by private individuals and the community as a whole.

A majority of the ST people continue to depend on underdeveloped traditional shifting agriculture for their livelihood that results in low productivity in comparison with wet land agricultural production. For example, the jhum foodgrains productivity is only 1050 kg/ha in 2014–2015 in Tripura (DES, 2015) as compared to the overall foodgrains productivity of 2620 kg/ha in 2011–2012 (Department of Agriculture and Co-operation, 2015). As such, under shifting cultivation in particular, the concept of surplus for trade hardly arises as the economy is based on self-subsistence (Das, 2006). Thus, generally, agricultural or food production is inadequate to meet the region's requirements (Sachdeva, 2000; Hussain, 2004; Husain, 2006; Marchang, 2016). It is clear that agricultural production is not even sufficient to be called subsistence because the region contributes only 3 per cent to India's foodgrains production (Marchang, 2016) against the region's population and geographical area contribution of 3.8 per cent and 8.0 per cent respectively in 2011.

The livelihood system of STs and their economy are gradually integrating and orienting towards the mainstream market economy. For example, in Tripura, the cultivation system has changed from shifting cultivation to capital intensive settled cultivation and the agricultural produce system has changed from family consumption production to commercialisation (Sengupta, 2013). Interestingly, most of the ST families have multiple sources of livelihood (Nongbri, 1999) and, their economic activities have been diversified into different occupations (Corbridge, 1988). Moreover, education has driven them, particularly the younger generation, to wage employment specifically for organised employment (Marchang, 2016) which indicates an upward occupational mobility in terms of stable salaried income and some social gains in terms of the prestige and status of being a salaried person. Labour mobility from rural areas as well as semi-urban areas comprising unskilled and semi-skilled labourers to urban areas where non-agricultural economic opportunities, associated with some status or prestige, are available is evident among the ST population. Similar is the situation where Foote and Hatt (1953) concluded that occupational movement was toward jobs of higher prestige.

The employment structure has changed from the agriculture sector, orienting towards the non-agriculture sector with changes in specialisation, technology, education and economy among others. The employment cross classified by broad industry as cultivators, agricultural labourers, household manufacturing and other workers in rural and urban areas among the STs of NE states and the country is presented in Table 10.3. A large section of the workers still depends on the informal sector, especially in agriculture. As Papola (1981) already noted, the size of the informal sector remains large in the region due to underdeveloped industrialisation. Agriculture remains the

Table 10.3 Share (%) of workers (main + marginal) by industry for ST in NE states/NER/India

Area	*State/ region/ country*	*Cultivators*		*AL*		*HHI*		*Other workers*	
		2001	*2011*	*2001*	*2011*	*2001*	*2011*	*2001*	*2011*
Rural	Ar.P	81.5	74.4	2.4	4.5	1.2	1.3	15.0	19.8
	Assam	69.7	61.6	11.9	15.7	3.7	3.9	14.7	18.8
	Manipur	66.9	67.2	7.5	7.4	5.7	3.4	19.9	22.0
	Meghalaya	58.8	52.5	20.1	20.1	2.2	1.9	18.8	25.6
	Mizoram	83.8	77.4	3.7	7.3	1.1	1.0	11.5	14.3
	Nagaland	77.0	71.3	3.7	6.6	2.5	2.1	16.7	19.9
	Sikkim	60.2	53.7	5.8	10.9	1.4	1.5	32.5	33.9
	Tripura	42.4	33.8	36.9	38.1	2.2	2.2	18.5	25.9
	NER	67.5	60.4	13.0	15.5	2.9	2.7	16.6	21.4
	India	47.1	36.9	38.4	47.1	2.1	1.7	12.5	14.3
Urban	Ar.P	11.4	7.7	1.9	2.2	1.4	1.6	85.3	88.5
	Assam	8.6	9.9	2.1	4.0	3.0	3.1	86.3	83.0
	Manipur	7.1	19.1	3.6	3.3	5.5	4.5	83.7	73.1
	Meghalaya	6.3	5.2	6.9	4.1	1.6	0.9	85.3	89.8
	Mizoram	25.3	17.2	8.7	10.1	2.1	2.2	63.9	70.5
	Nagaland	7.3	14.0	1.2	4.3	3.9	3.4	87.6	78.3
	Sikkim	0.2	1.8	0.3	2.1	1.0	1.4	98.4	94.7
	Tripura	1.7	6.2	2.6	11.0	0.9	1.1	94.8	81.8
	NER	15.0	12.1	5.9	5.9	2.3	2.3	76.7	79.6
	India	6.5	5.8	12.4	13.3	2.9	2.5	78.2	78.3
Total	Ar.P	75.0	66.1	2.3	4.2	1.2	1.3	21.5	28.4
	Assam	67.7	59.3	11.6	15.2	3.7	3.8	17.0	21.7
	Manipur	64.9	62.6	7.4	7.0	5.7	3.5	22.0	26.9
	Meghalaya	52.9	46.0	18.6	17.9	2.1	1.7	26.4	34.4
	Mizoram	58.1	49.4	5.9	8.6	1.5	1.6	34.5	40.4
	Nagaland	71.4	61.9	3.5	6.2	2.6	2.3	22.4	29.5
	Sikkim	56.1	45.7	5.4	9.5	1.4	1.5	37.0	43.2
	Tripura	41.6	32.8	36.2	37.2	2.1	2.2	20.0	27.8
	NER	62.4	54.3	12.3	14.3	2.8	2.6	22.5	28.7
	India	44.7	34.5	36.9	44.5	2.1	1.8	16.3	19.2

Source: Author's calculation based on RGCCI (Table ST2 and ST4, 2001 and PCA, 2011).

predominant avenue of employment and source of income for rural people in particular in NER, as in the country. This corroborates with the household type or economic group condition as discussed earlier. The share of cultivators was much larger in the region than in India, perhaps determined by the difference in the landholdings and its ownership system in the hill areas of NER and the ST areas of the country. STs are ruled by certain distinct characteristics ranging from the community landholding system to accessing land. Since land in rural areas of NE is largely owned by the community and some by individuals (Maithani, 1991, 2005; Sachchidananda, 1989) the incidence of landlessness is negligible (Maithani, 1991). Nevertheless, there is no uniform land tenure system across the tribes in the region (Marchang, 2016).

NSSO (2001, 2015) data on land holdings and land cultivated areas, Marchang (2016) shows that land is underutilised for agricultural purpose in the region when compared to the national level. The size of land holdings, which determines the agricultural intensity and agricultural income, among the ST households is small and medium in size; that is perhaps a private or individual land. The size of land holdings varies across the NE states. ST dominated states like Arunachal Pradesh and Nagaland have a larger proportion of households holding a relatively bigger size of land of above 2 hectares per household. Marchang (2016, 2017) showed that in NE, a large share of the households cultivated small sizes of land which resulted in low per capita agricultural production and subsistence production.

In India, a majority of agricultural labourers (AL) are drawn from the Scheduled Castes and Tribes (Sethi, 1984). This is delusive for the region as the share of AL is abysmally low in NER for India. The situation is likely because of the prevailing unique land tenure system, particularly community land ownership in the ST regions. Further, employment in household manufacturing industry (HMI) continues to be low in the region; however, slightly higher in the region than the country as a whole. Even the share of other workers remains larger in the region.

Data shows that employment in the agriculture sector is lower in the region than in India (as a whole) or in other words, non-agricultural employment is more prominent in the region than in the country (Table 10.3). In 2011, about 54 per cent of the total workers in the region were engaged as cultivators, against the national average of about 35 per cent. Employment in cultivation has declined as the share of cultivators decreased in the region, following the national trend, during 2001–2011. Over the years, the share of AL has increased slightly by about two percentage points in the region following the national trend of increase by about eight percentage points. The share of household of manufacturing industrial workers also declined notably. The share of "other" category workers has increased by about six percentage points in the region when it increases by about three percentage points in India during 2001–2011. This change in the structure of employment is a result of rapid structural changes in the economy. As Stewart (1987) concluded, rapid structural changes are accompanied by occupational transition in which employment shifts from agriculture to non-agriculture urban pursuits, informal to formal sector and traditional to modern industries and services. A similar pattern and trend prevails for the NE states as well. In 2011, among the NE states, Arunachal Pradesh had the highest share of cultivators with about 66 per cent followed by Manipur, Nagaland and so on, and the least was in Tripura with about 33 per cent. Arunachal Pradesh, Manipur and Nagaland had a higher share of cultivators than the region in the same year. It was higher in all the NE states, except Tripura, when compared to the country as a whole. The share of AL ranges between the lowest in Arunachal Pradesh with about 4 per cent and the highest in Tripura with 37 per cent. It was higher than the regional figures only in the state of Assam, Meghalaya and Tripura. Interestingly, none of the NE states has a higher share of AL than the rest of the country with a share of about 45 per cent.[10] It was significantly low among the ST dominated states such as Arunachal Pradesh, Mizoram and Nagaland. It is even low in the state of Manipur where a quarter of the population are STs and more than 90 per cent of lands are owned by them.

Employment in the household manufacturing industry (HMI) was not very prominent in most of the NE states, particularly Sikkim and ST dominated states like Arunachal Pradesh, Meghalaya and Mizoram. The share of workers in HMI was the lowest in Arunachal Pradesh with about 1 per cent and highest in Assam with a share of close to 4 per cent in 2011. Assam and Manipur were the two NE states which have a higher share of it than the region as a whole. It was higher in Assam, Manipur, Nagaland and Tripura when compared to the country's level of 1.8 per cent in the same year. In the case of employment in the others category of work, the share ranges from about 22 per cent in Assam to 43 per cent in Sikkim. Meghalaya, Mizoram, Nagaland and Sikkim have a higher share of the others category of workers when compared to the region's share of about 29 per cent. Interestingly, its share was higher in all the NE states than India's share of about 19 per cent.

A rural and urban comparison shows that in general, the employment in rural areas was dominated by agricultural employment, i.e. both cultivators and AL taken together, while urban employment was dominated by non-agricultural employment in all the NE states and the region following the national pattern in both years. Nevertheless, the means of livelihood has converged towards non-agricultural activities or employment as the level of education rapidly improves. Marchang (2016) using RGCCI, i.e. Census of India (2011) and NSSO (2015) showed that literacy rates for NE states far exceed the country's level among the STs. Also, literacy rates are exceedingly higher for STs of NE states when compared with all social groups of India. Sundaram and Tendulkar (2003) opined that lack of exposure to education and isolated habitation from the social mainstream made STs vulnerable to exploitation by non-STs. The connotation of geographical isolation appears because of the lack of developmental characteristics such as roads in the sparsely populated hilly region. The tribal people in the hills generally lack education, training, communication facilities, etc. (Singh, 2007). This is partially incorrect because of the extent of educational development in NER. Both decennial censuses of India and NSSO data showed a rapid educational growth systematically for STs, partly due to an increase of educational enrolment ratio (Marchang, 2016). It is also noteworthy that some tribes have transgressed on others educationally (Goswami, 1984).

Similarly, among the Naga tribes of Manipur, the Tangkhuls who embraced Christianity first outperformed in education as compared to other tribes (Marchang, 2013).

The change in the structure of employment and income, from agriculture to non-agriculture, is attributed to the development of education and improvement of the industrial sector (Marchang, 2016). Sengupta (1988) pointed out that inappropriate development strategies have not brought any relief to the STs. The STs have been uprooted in the course of industrial expansion and have been discriminated against in job opportunities. However, Singha (2011) asserts that in Manipur, they have developed significantly due to the growth of English education imparted by Christian missionaries.

Occupational changes and livelihood status of Manipur

Analyses of primary field data from Manipur shows that STs owned largely inherited land, especially in rural areas. They owned land in the hill slopes, terraces and in the plains. Some of them have land both in the hills and/or terraces and/or plains. Rural people largely own land for agricultural use. Very few households do not have land. Most of them are small land owners (less than 1 hectare per household).[11] Most of the agriculturists, as their primary or secondary occupation, cultivated about 1 hectare and as a result, the agricultural produce is subsistence production. Urban people and/or non-agricultural workers were also pursuing agricultural activities, as a secondary source of income, for their livelihood. Cultivation is predominant among the lower income households.

Primary field data results validate that the means of livelihood of STs have undergone a change. Self-employment (62 per cent), particularly in agriculture activities, remains a prominent source of livelihood, especially in rural areas. The self-employed include cultivators, farmers, carpenters, drivers, businesspersons (shopkeepers, vegetable vendors, entrepreneurs, local wine brewers, etc.), tailors, weavers, social workers and others. As much as 83 per cent of the self-employed are cultivators and farmers. However, regular salaried employment is the main source of livelihood for a majority of the urban people. Overall, most of the ST households depend on a non-salaried income. Cultivation, as a secondary activity, is visible among the non-agriculturist households such as regular salaried households. It appears that many households have multiple or mixed sources of income. The type of cultivation includes shifting, terrace and permanent. Also, some households practise a combination of these three types of cultivation such as shifting and terrace or shifting and permanent or terrace and permanent. Most of the households belong to low and medium income households (monthly income below Rs 15,000). These households spent as much as and even more (by borrowing) than their monthly income; thereby, indebtedness is prevalent among them. This portrays a subsistence income irrespective of the sources of income. The higher income group households spent relatively less than their monthly income, so some savings are accrued. Non-agricultural households such as salaried households exhibit a higher living standard and have a greater potential and tendency to save.

Conclusion

The means of livelihood of the ST population has changed from agricultural to non-agricultural. It is evident from three facts: decline of agricultural households while non-agricultural households increase; decline of agricultural income while non-agricultural income rises; and shift of employment from agricultural to non-agricultural activities. However, agriculture, shifting cultivation in particular, continues to be a prominent means of livelihood for some of them. Their means of livelihood has changed from subsistence agricultural income towards diversified modern market-oriented employment and economy. The change of livelihood means manifested by the employment and economy orienting towards non-agricultural sector is essentially due to education and economic mobility and vice versa. This change is associated with a systematic increase of per capita income and educational levels.

Notes

1 This chapter is a modified version of the working paper published by the Institute for Social and Economic Change, Bengaluru in March 2019
2 It consists of Arunachal Pradesh, Assam, Manipur, Meghalaya, Mizoram, Nagaland, Sikkim and Tripura.
3 As per the 2011 Census of India, there are 241 constitutionally recognised STs in NER. The ST population constitutes 8.61 per cent of the total population of India in 2011. About 31 per cent of the ST population of the region was in Assam and the rest were in Ar.P (7.7 per cent), Manipur (7.3 per cent), Meghalaya (20.6 per cent), Mizoram (8.4 per cent), Nagaland (13.8 per cent), Sikkim (1.7 per cent) and Tripura (9.4 per cent).
4 The chapter is partly based on a project entitled *Impact of education and employment on the economy of scheduled tribes in north east India* sponsored by the Indian Council of Social Science Research, New Delhi (2016).
5 A household is a unit where the members of a family share a common kitchen. The reference period of the survey was the preceding one year from the date of survey.
6 Also attributed by MGNREGA's 100-day guaranteed unskilled rural work per year to each rural household.
7 States' income is considered as proxy for ST income since a separate data for STs is unavailable. States dominated by ST population would reflect, represent and interface the income situation.
8 In 2012–2013, only 1.9 per cent of India's 2.2 lakh factories were in NER and the region contributed only 1 per cent value of factories output in the Rs 60 lakh crore output of India as a whole (Marchang, 2016).

9 Area under shifting cultivation declined from 5848 km^2 in 2003 to 3924 km^2 in 2009 in NER. Number of shifting cultivating families increased from 684,000 in 2001 to 748,000 in 2011 in NER. See Marchang (2016) for details.

10 Perhaps partly due to the nature of land ownership, i.e. community in most of the NE states, especially in the ST dominated areas; partly due to equitable land distribution; and partly because of low density of population.

11 Secondary data has also shown a similar result (Marchang, 2016).

References

Bhagat, R. B. (2013). Conditions of SC/ST households: A story of unequal improvement. *Economic and Political Weekly*, *48*(41), 62–66.

Chaube, S. K. (1999). The scheduled tribes and Christianity in India. *Economic and Political Weekly*, *34*(9), 524–526.

Christoph, von F. (1982). *Tribes of India: The struggle for survival*. Berkeley, CA: University of California Press.

Corbridge, S. (1988). The ideology of tribal economy and society: Politics in the Jharkhand 1950–1980. *Modern Asian Studies*, *22*(1), 1–4.

CSO. (2013). *State domestic product (state series)*. New Delhi: MOSPI. Retrieved November 25, 2013, from http://mospi.nic.in/Mospi_New/admin/publication.aspx

Das, D. (2006). Demystifying the myth of shifting cultivation: Agronomy in the north-east. *Economic and Political Weekly*, *41*(47), 4912–4917.

Department of Agriculture and Co-operation. (2015). Statewise Area, Production and Yield Statistics – 10 year ending (1996/97 to 2011/12), Dept of Agriculture and Co-operation, GOI. Retrieved March 13, 2015, from http://eands.dacnet.nic.in/APY_96_To_07.htm

DES. (2015). *Tripura at a glance – 2014–15*. Agartala: Planning (Statistics) Department.

Foote, N. N., & Hatt, P. K. (1953). Social mobility and economic advancement. *The American Economic Review*, *43*(2), 364–378.

Goswami, A. (1984). Tribal development with special reference to north-east India. *Social Scientist*, *12*(8), 55–60.

Gupta, A. K. (2000). Shifting cultivation and conservation of biological diversity in Tripura, northeast India. *Human Ecology*, *28*(4), 605–629.

Hanumantha, P. R., & Grover, D. (1979). Employment planning for scheduled castes and scheduled tribes. *Economic and Political Weekly*, *14*(24), 1015–1022.

Horan, P. M. (1974). The structure of occupational mobility: Conceptualization and analysis. *Social Forces*, *53*(1), 33–45.

Husain, Z. (2006). Strategy for food and nutrition in north-east India. In D. Basu, B. K. Francis, & B. D. Ray (Eds.), *Agriculture, food security, nutrition and health in north east India* (pp. 81–90). New Delhi: Mittal Publication.

Hussain, M. (2004). Food security and the north-east. *Economic and Political Weekly*, *39*(41), 4515–4516.

Kumar, A., & Ramakrishnan, P. S. (1990). Energy flow through an Apatani village ecosystem of Arunachal Pradesh in northeast India. *Human Ecology*, *18*(3), 315–336.

Kumar, A. G. (1987). The evolving pattern of the economy of Tripura. In V. S. Mahajan (Ed.), *Emerging pattern of north-eastern economy* (pp. 246–251). New Delhi: Deep and Deep Publication.

Maithani, B. P. (1991). Rural development in north-east: Problems and prospects. In R. K. Samanta (Ed.), *Rural development in north-east India: Perspectives, issues and experiences* (pp. 59–68). New Delhi: Uppal Publishing House.

Maithani, B. P. (2005). *Shifting cultivation in north east India: Policies, issues and options*. New Delhi: Mittal Publication.

Marchang, R. (2013). Population and educational development of Nagas: A brief note. In P. C. Dutta & K. Singha (Eds.), *Ethnicity, resources and institution for development of north eastern states of India* (pp. 188–212). New Delhi: Akansha Publishing House.

Marchang, R. (2016). *Impact of education and employment on the economy of scheduled tribes in north east India* (Unpublished report). New Delhi: ICSSR.

Marchang, R. (2017). Land, agriculture and livelihood of scheduled tribes in north-east India. *Journal of Land and Rural Studies*, *6*(1), 67–84.

MEF & GBPIHED. (2009). *Governance for sustaining Himalayan ecosystem guidelines & best practices*. New Delhi: Ministry of Environment & Forests and G.B. Pant Institute of Himalayan Environment & Development.

MTA. (2013). *Statistical profile of scheduled tribes in India 2013*. New Delhi: Ministry of Tribal Affairs, Statistics Division, GOI.

Nongbri, T. (1999). Forest policy in north-east India. *Indian Anthropologist*, *29*(2), 1–36.

NSSO. (2001). *Employment and unemployment situation among social groups in India 1999–2000*. Report No. 469(55/10/7). New Delhi: MOSPI.

NSSO. (2015). *Employment and unemployment situation among social groups in India*. Report No. 563(68/10/4). New Delhi: MOSPI.

Papola, T. S. (1981). *Urban informal sector in developing economy*. New Delhi: Vikas Publishing House.

Peale, S. E. (1874). The Nagas and neighbouring tribes. *The Journal of the Anthropological Institute of Great Britain and Ireland*, *3*, 476–481.

Reddy, B. S. N. (1991). Shifting cultivators and rural development in Garo hill: Some reflections. In R. K. Samanta (Ed.), *Rural development in north-east India: Perspectives, issues and experiences* (pp. 216–223). New Delhi: Uppal Publishing House.

RGCCI. (2001/2011). *Census of India*. New Delhi: Registrar General and Census Commissioner of India.

Roy, B. B. K. (1989). Problems and prospects of tribal development in north-east India. *Economic and Political Weekly*, *24*(13), 693–697.

Roy, R. D., Xavier, B., & William, M. S. (2012). Study on shifting cultivation and the socio-cultural integrity of indigenous peoples. *Economic and Social Council*. New York: United Nations. Retrieved February 20, 2015, from http://daccess-dds-ny.un.org/doc/UNDOC/GEN/N12/241/85/PDF/N1224185.pdf?OpenElement

Sachchidananda. (1989). *Shifting cultivation in India*. New Delhi: Concept Publishing Company.

Sachdeva, G. (2000). *Economy of the north-east: Policy, present conditions and future possibilities*. New Delhi: Konark Publishers Pvt. Ltd.

Saikia, J. (1991). People's participation in rural development: A study in Nagaland. In R. K. Samanta (Ed.), *Rural development in north-east India: Perspectives, issues and experiences* (pp. 186–195). New Delhi: Uppal Publishing House.

Seavoy, R. E. (1973). The shading cycle in shifting cultivation. *Annals of the Association of American Geographers*, *63*(4), 522–528.

Sengupta, M. (2013). Shifting cultivation and the Reang tribe in Tripura. *Economic and Political Weekly*, *48*(40), 56–65.

Sengupta, N. (1988). Reappraising tribal movements III: The economic basis. *Economic and Political Weekly*, *23*(21), 1054–1055.

Sethi, R. M. (1984). Changing patterns of female labour in agriculture: The case of Punjab. *Social Action*, *34*(4), 354–367.

Shimray, U. A. (2004). Women's work in Naga society: Household work, workforce participation and division of labour. *Economic and Political Weekly*, *39*(17), 1698–1711.

Sikidar, S. (1990). *Economic development of tribal India*. New Delhi: Ashish Publishing House.

Singh, T. B. (2007). Poverty in Manipur. *Economic and Political Weekly*, *42*(3), 251–254.

Singha, K. (2011). *Economy of a primitive tribal village: A study of Maram in Manipur*. New Delhi: Concept Publishing Company.

Srivastava, V. K. (2008). Concept of 'tribe' in the draft national tribal policy. *Economic and Political Weekly*, *43*(50), 29–35.

Stevens, P., & Weale, M. (2003). *Education and economic growth* (pp. 1–28). London: National Institute of Economic and Social Research.

Stewart, C. T., Jr. (1987). Structural change and intergenerational occupational mobility. *The Journal of Developing Areas*, *21*(2), 141–158.

Sundaram, K., & Tendulkar, S. D. (2003). Poverty among social and economic groups in India in 1990s. *Economic and Political Weekly*, *38*(50), 5263–5276.

Thangchungnunga. (1987). The agricultural profile in Mizoram. In V. S. Mahajan (Ed.), *Emerging pattern of north-eastern economy* (pp. 268–273). New Delhi: Deep and Deep Publication.

Vaizey, J. (1967). Education and the Irish economy. *The Irish Journal of Education/Iris Eireannach an Oideachais*, *1*(2), 113–123.

11

"OUTSOURCING" IN COAL MINING

Understanding labour, livelihood and mafia politics in the coalfields of Dhanbad, Jharkhand

Sujit Kumar

Introduction

Rise in inequality, regional imbalances in economic growth and compromise with labour welfare (Papola & Sharma, 2004; Sanchez, 2010, 2012) are a few macro implications of the economic reform policies in India. Among others, labour has been one of the most crucial and controversial aspects of liberalisation programmes in India. Even before liberalisation, a large part of labour existed in the informal sector and faced the uncertainties related to wages, working hours, working conditions and tenure and so on. But in the post-liberalisation era, even the sectors which ensured formal employment have adopted measures like contractualisation, retrenchment and outsourcing in the name of restructuring.[1] Scholars have attempted to study the various implications of labour informalisation. Interestingly, many studies have found that informalisation is motivated by the quest to *discipline* labour and generate profit leading to the decline of labour welfarism. This has brought consequent changes in the nature of labour politics wherein the target of protest has shifted from employers to the state and the regime has shifted from "work to welfare" (Agarwala, 2006). While the state has come to ensure the role of the employer in this framework, what will happen if the state itself disowns its pledge towards labour welfare? This question becomes the vantage point of our inquiry which is conducted through the study of a public sector unit (PSU), which is Bharat Coking Coal Ltd. (BCCL) engaged in coal mining in Dhanbad, Jharkhand.

BCCL, since 2005, has started outsourcing some of its production related activities. The company has been incurring losses for the last several years in its mines located in certain areas of Dhanbad. The poor performance of the BCCL can be attributed to the poor managerial skills of the executives as well as the nature of labour politics. However, the government has responded with strategic measures like outsourcing to generate profit from the enterprise. While the strategy has visibly failed to yield the desired results, it has put the labourers at the receiving end by drastically curtailing the prospect of labour politics.

The present chapter intends to deconstruct the policy of "outsourcing", analyse the livelihood strategies of people in the coalfields and investigate the nature of the mafia–politics–state nexus. To be precise, our aim is to understand the policies like outsourcing as a government's business ethics and its implications for the different stakeholders. The chapter begins by analysing the policy of "outsourcing" as adopted by public enterprises engaged in coal mining activities from three perspectives. The first is from a government entrepreneurial approach wherein "outsourcing" is endorsed as a policy to increase production and generate profit. Here we will discuss the findings pertaining to the outsourced blocks in South Bhowra Colliery. The second perspective is to capture the nature of the political economy of outsourcing by analysing the dual role played by the violent actors, regarded as coal *mafia*, both as trade union leaders and state legislators. And the third perspective captures the trajectory of livelihood options available to the different groups of people, particularly the Santhal adivasis.

The discussions are based upon a primary survey carried out between May 2015 and May 2017 in Dhanbad district of Jharkhand. In-depth interviews were conducted with Coal India executives, local politicians, trade union leaders from diverse organisations and contractual workers. Besides this, field trips to several villages were undertaken in order to understand the nature of the livelihood and their intimate relations with coal mining in the areas. The villagers interviewed were farmers, fishermen and workers in coal mines.

A prelude to outsourcing in the coal sector

If one is asked to provide convincing reasons for the nationalisation of the coal sector, one will certainly hesitate to express any reason except one, i.e. for energy security. The other reasons cited for nationalising the sector have been evaluated and criticised vigorously.[2] For instance, while rubbishing most of the reasons, Kumar (1981) criticised the government for revising the estimate suggested by the Fuel Policy Committee under Sukhmoy Chakraborty[3] and observed that the arguments endorsed for nationalisation were insufficient and some of the problems like separate pricing of coking and non-coking coal continued even after nationalisation. However, his analyses did not take into account the various domestic and international factors influencing not only the production but also the demand for coal throughout the 1970s. Another limitation in Kumar's analysis is the sheer absence of the political economy approach before he arrives at the conclusion that the pricing of coal being controlled by the Coal Price Revision Committee (CPRC) has failed to take into account the increased prices of mechanisation, thus discouraging the private sector. For instance, Subhomoy Bhattacharjee (2017, p. 54) claims that due to the coal price control imposed through the Colliery Control Order of 1944, the private miners suffered a loss on every tonne of production as the price was fixed at Rs 42/tonne while it sold in the black market at Rs 100/tonne. The price control was rescinded only in 2000. Hence, even during the period of nationalisation, coal could not be sold by the CIL at market-determined prices as most of its consumers were other public sector units and a rise in price would have adversely affected them and, further, the end users.

After nationalisation, however, the coal trading companies became authorised coal handlers and the strongmen employed by the private miners became the trade union leaders in the state-run companies. Thus, as scholars observed (see Bhattacharjee, 2017, p. 7) the shift to nationalisation has groomed the political mafia leaders, a theme to which we will return later in the chapter. Throughout nationalisation, a blame game continued for the declining coal output towards the end of which a disproportionate responsibility was fixed on the labour and its indiscipline, absolving the other factors, viz. corrupt management and politically connected mafia in the coalfields. Sengupta (1999, p. 29) identifies the pre-liberalisation structure of the industry, infrastructure bottlenecks and statutory and legal obstructions in the form of environmental, labour, mining and coal governing laws as the factors which retard competition in the short and medium run. In line with its policy of liberalisation, the Government of India constituted a Committee on Integrated Coal Policy (CICP) in 1995 which submitted its report before Parliament in May 1996. Some of the major recommendations were: to amend the Coal Mines (Nationalisation) Act, 1973 allowing any Indian company to mine coal and lignite not only for captive consumption but also sale; allow foreign investors to set up 100 per cent subsidiaries for mining and exploration; competitive bidding of coal blocks on the criteria of minimum recoverable percentages and minimum transfer price to both nationalised coal companies and private companies. Simultaneously, during the second leg of deregulation in February 1997, the central government proposed to amend the Mines and Minerals (Regulation and Development) Act of 1957 to set up an independent body to allocate blocks on the basis of competitive bidding. Already, the expert group on commercialisation of infrastructure projects constituted under Rakesh Mohan in 1994 had recommended an independent regulatory body for coal. The proposed body was supposed to perform the threefold task of regulation, facilitation and adjudication (Sengupta, 1999, p. 32). Independent of government, the quasi-judicial regulatory body was proposed to be accountable to the judiciary (ibid.).

Given the view that even in 2047 – the centenary of Indian Independence – India will depend upon fossil fuel for almost two-thirds of its energy requirements, one can hardly ignore coal as it will fulfil half of this estimated requirement (IEA, 2015). However, despite the presence of vast reserves, India will have to import almost one-third of the coal to be burnt in the steel industry or thermal power plants. But even import provides a limited solution as the ports facilitating the import reach optimum levels of handling. This makes the extraction of domestic metallurgical and non-metallurgical grades of coal imperative. The generally accepted bottlenecks behind this deficiency in domestic production are: ownership of almost 80 per cent of the estimated coal reserve with Coal India Ltd.; stringent land acquisition and environmental laws; absence of a transparent regulatory framework, etc. Particularly, the last reason created havoc in the political discourse of India as the CAG in the year 2012 discovered a loss to the tune of Rs 1.86 lakh crore to the exchequer due to misallocation of coal blocks. At the broader level, people have suggested the adoption of the Mining, Development and Operation (MDO) model under which the private sector will be allowed to undertake mining while the CIL will own and sell the coal (Khanna, 2013, p. 22). Even "outsourcing" is considered as a reasonable strategy even though it was disallowed through a notification in 1975 prohibiting contractualisation of raising and loading coal as well as clearing of overburden during open cast mining (Roy, 2003, p. 5154). Roy states that "the same rule still continues, but now the government has started a contractual system without changing the rule, prohibiting it and creating a ridiculous situation where the Regional Labour Commissioner (central), a government servant, is prosecuting the government without effect". Roy (2003, p. 5154) further argues that the policy of "outsourcing" is mostly promoted by the managers of the public sector as it helps them to project the "success" of nationalisation.

Another aspect of outsourcing is related to the nature of mining as all of them happen to be Open Cast Projects (OCPs)[4]. It is worth mentioning here that underground mining, though expensive and sophisticated, has proved to be environment-friendly and holds a better prospect of capacity utilisation. Despite these virtues, the underground mines are being closed on grounds of being uneconomic. However, Kuntala Lahiri-Dutt (1999, p. 2955) who has extensively studied the coal mining sector, considers the underground mining to be the *inevitable future* of the coal industry. In her case study of the Raniganj underground mines under Eastern Coalfield Limited (ECL) focused on understanding its sickness, she arrives at the conclusion that the mines became unviable because of the additional *coal cess* imposed by the West Bengal government, making the production dearer by Rs 135 per tonne (ibid.). Thus, it is not very difficult to understand that whatever profit and increase in production is noticed through outsourcing is primarily because of reckless OCPs and the money which BCCL saves by avoiding direct hiring of manpower.

Outsourcing: more than making the enterprise profitable!

The BCCL was formed on 1 May 1972 when around 220 privately run coking coal mines were nationalised. By 1973, the government of India took over the non-coking coal mines also and finally the Coal India Limited (CIL) was formed in 1975 to manage all the nationalised mines. BCCL's policy of outsourcing needs to be understood against the changed policy-making context in India since 1991. Being a loss-making unit but crucial for the energy requirements of the country, the government has decided to revive the coal sector rather than disinvesting it. The impetus for increased production is thus provided through auctioning of new coal blocks as well as outsourcing the production in loss-making units. This method of reviving a "sick PSU" has direct implications as it increases the informal workforce in the organised sector. The other dire consequences are death of labour politics and deterioration of employment conditions for the workers. For instance, as studies have already revealed, in the name of labour flexibility, informalisation has weakened the position of trade unions as the workers, if fortunate, are under firm level unions (Papola & Sharma, 2004; Agarwala, 2006). Similarly, the National Commission for Enterprises in the Unorganised Sector (NCEUS) also observed that contract labour has grown significantly in manufacturing and services sectors with a rise in wage differentials. All this looks contradictory as India is also a party to the recommendation adopted by the International Labour Organisation (ILO) in 2015 which says that "governments should facilitate the transition of workers from the informal sector to the formal sector" (Rajalakshmi, 2016, p. 36). In short, instead of working towards a commitment to improve labour conditions, the PSUs have actively pursued policies like outsourcing. In fact, the government has further amended the Factories Act, relaxing the labour laws for small and medium units employing up to 40 workers. Labour being a concurrent subject, various state governments have also amended the Factories Act, the Industrial Disputes Act and the Trade Union Act leading to compromise with labour interests (ibid.). It is against this general picture that we will try to understand the emerging situation in BCCL.

The nationalisation drive was motivated by the state's quest to develop resources according to the country's growing energy demand and ensure better working conditions to those employed in the coal mines. In fact, the nationalised coal mines gave further impetus to the rise of semi-urban centres which were deprived of their old identities, parochial roots and feudal characteristics. These habitations had to be now occupied with people who pledged to contribute in the *making* of modern India. Apparently, things progressed well and the meeting point of diverse cultures gave rise to a fluid cultural notion with unique plural identity which made sense when lived in its plurality but appeared defaced when attempted to provide an "exclusive" meaning. Maybe, the "unity in diversity" of Nehru in post-Independence India was to materialise in the vicinity of "temples of modern India". Disregarding this contemplation is again very difficult. The cosmopolitan culture and behaviour is typical of many such small towns which mushroomed due to industrialisation. Even if these centres are not as affluent as a metropolis, they displayed a unique cultural feature which many would identify as modern Indian culture.[5] One can go to some lengths in exploring the difference such societies posit in comparison to the traditional societies. But our concern here is to understand the policy of "outsourcing" as adopted by PSUs like BCCL.

In the first phase of adopting outsourcing, the BCCL hired Macintosh Burn Ltd in the year 2005 to mine the coking coal stuck under the abandoned railway track between Dhanbad and Patherdih. Subsequently, the company has outsourced many of its production-related activities to the contractors, thus resulting in a steady rise in the number of contractual workers. This trend is in conformity with the general picture of rise in informal workers within the organised sector. Outsourcing in the South Bhowra Colliery started since 2010 and two companies Demo and SAR Parivahan worked between 2010 and 2015. However, A T Deo Co, owned by Lal Babu Singh, a close aide of Jharia MLA Sanjeev Singh, is at present working in the only outsourced block. BCCL provides vocational training and conducts medical fitness tests for the workers hired by the contractors. A total of 74 workers of both skilled and semi-skilled nature are hired by the contractor. Managerial and supervisory staff from BCCL conduct the blasting of coal seams using explosives and oversee the work on these sites. The semi-skilled and skilled workers largely employed in operating the heavy machines are mostly from the families of the government employees working in the mines as manual workers. Ironically, even the vocational training received by their wards has not been able to ensure a gainful employment. The compulsion of securing a livelihood and the absence of other remunerative options have made these youths subject themselves to such exploitation. Sometimes, the prospect of regularisation under the PSUs acts as a motivation for these youths to continue working under contract. However, only in rare cases do we find that they managed to secure a government job which is again contingent upon their political clichés.

Mafia, politics and outsourcing

This section attempts to understand the framework within which the labour issues need to be located in the mainstream political discourse of Dhanbad. It, hence, explores the interaction between trade unions, politics and violence in the city. In her study of the jute mills in and around Kolkata, Nandini Gooptu (2007) provides a framework to understand and locate the receding space for labour politics based on class owing to the rise of identity-based politics. To capture the shift in labour politics through this approach is linear in nature and proves to be useful where the discourse has shifted from "work to welfare" (Agarwala, 2006). However, to understand the discourse of labour politics in the coalfields of Dhanbad, the use of a "dyadic" approach is proposed here. According to this approach, identity was typically a part of trade union politics in Jharkhand along with class. We will explore this relationship between identity, trade unionism and class by juxtaposing the trade union politics of the left-affiliated unions with that of the trade unions of the political parties like BJP, Congress and Janata Party. At the superficial level, the ethnic issues might look like a part of labour politics in the state-run enterprise. But a close analysis reveals its embedded geographical conflict as the left-affiliated trade unions were largely supported by the Bengalis and the Jharkhandis who mostly worked as manual labourers in the mines. However, the trade unions affiliated to parties like the BJP, Congress or the Janata Party were dominated by the people of Bihar and Uttar Pradesh.

Conventionally, trade unionism in India is divided into four phases[6] (Kalarivayil & Nair, 2018). The first phase is "largely associated with the expansion of organised sector along with high growth in public sector employment" (ibid.). During the second phase, one can observe a rise "of different micro movements against the failed promises of the state" with a high level of industrial strife. Widespread unionisation and industrial strife in both public and private sectors characterised by intense economic and political crisis marked the third phase, while the fourth phase is identified with a shift in the state's role focused at managing labour relations through policy and legal instruments (ibid.). In the post-reform period, there is a decline in the membership of formal workers across Central Trade Union Offices (CTUOs) affiliated to national political parties. However, the trade unions responded "to the realignment of state, capital and working class in the neoliberal period by building partnerships and broad coalitions to achieve social policy goals" (Mohanty, 2009 as quoted in Kalarivayil & Nair, 2018, p. 53). As a part of this strategy, the New Trade Union Initiative (NTUI) formed in 2006 is firmly grounded on three principles – *unity, democracy and militancy*. In fact, two developments during the 1980s and 1990s prepared the ground for this initiative; first was the rise of independent unions at the local workplace level in the private sector, the second being the labour flexibility after the liberalisation of economy (Bhattacharjee & Azcarate, 2006, p. 65). While there are many success stories[7] of the NTUI, it is also gaining strength from its apolitical nature and strategic way of dealing with the employers and the state.

Though such an initiative has measured relevance in the unorganised and informal sector, it does not provide a protective cover and galvanise the contractual workers employed in the public sector or private sector. The contractual workers in the outsourced units of BCCL and Tata Iron and Steel Company (TISCO) are a living testimony to this conclusion. However, there exists a defining difference in the way the CTUOs interact with the contractual workers in these two cases. Whereas the voice of the contractual labour employed in BCCL units is presumably raised by the CTUOs, the only recognised trade union, Tata Workers Union (TWU), does not represent the contractual workers in TISCO. Analysing the role of violence in trade union politics in the context of the Jamshedpur Tata plant, Andrew Sanchez (2010, p. 169) argues that the casual workers employed by the company are not represented by the trade unions as the latter have been bribed by the company. However, the situation in the coalfields seems to differ very strongly. Being a public sector enterprise, the issue of "company bribing" is ruled out plain and simple. As against this, we argue that the powerful presence of criminal–trade union–politics nexus has given rise to a situation where the trade unions negotiate on behalf of casual workers for hike in wages. The trade unions raise the issue of wages without encouraging labour politics among the ranks of contractual workers. Honest trade union leaders like V. G. Gopal in the Tata Steel plant considered early retirement and mass casualisation as antithetical to the principles of collective action (Sanchez, 2010, p. 168). But the trade union leaders in the coalfield have also come to be the contractors for the outsourced workers in a certain number of cases. Hence, they negotiate for wage hikes as any hike of statutory wages by the company does not actually reach the contractual workers but is appropriated by the contractors.

Looking at the macro-level political scenario, one can easily capture this divergence within the labour politics as A. K. Roy, a communist leader who later played a key role in founding the Jharkhand Mukti Morcha (JMM) during the early 1970s, was posited against the likes of Suraj Deo Singh who earlier derived support from the Janata Party and later from the BJP. The role played by communist trade union leaders in the demand for a separate Jharkhand was also motivated by their interest to corner the Hindi-speaking people in the political discourse (Munda & Mullick, 2003). Nevertheless, a common phenomenon with both the types of trade unions was the rise of elites among their ranks. While the communist trade unions are still led by old-timers like S. K. Bakshi, the new generation of leaders in the union has acquired skills to negotiate with the changing political discourse. It is essential to capture this shift in order to understand the rise of mafia and violence in the labour politics of Dhanbad.

What seems to be particularly problematic in the context of PSUs in Dhanbad is the different dynamics of relationship between the state, violence and labour. While the framework of "enforcement partnership"[8] proposed by Vadim Volkov (2002) is applicable in the case of the private sector, the same cannot be used to understand the situation in the government sector. As explained above, while "company bribing" can be a strategy in the private sector, the same cannot be replicated in the PSUs. Mafia in Dhanbad has been historically linked with trade unionism in the PSUs and has come to benefit as contractors for the outsourced jobs. The political domination of the mafia and their simultaneous presence in trade unions provide them a certain edge over their counterparts in Jamshedpur who work for the company. Unlike them, the mafia in Dhanbad enjoys autonomy which is not only used to arm-twist BCCL occasionally as trade union leaders, but also ensure the smooth functioning of the contracted workers through their goons.

The discourse of violence is *systemic*[9] with wider implications for productivity, leading to profit for the state-run units. However, unlike Sanchez's (2016, p. 70) finding, in case of the Tata industry,[10] the systemic corruption discourse in the case of BCCL seems to placate the liberalisation tools like outsourcing while attending to the fear of the managerial class in the PSU and providing assured means of income to the long-embedded political *mafia* elements in Dhanbad. It is noteworthy that Janta Mazdoor Sangh (JMS) is the most powerful trade union in Jharia under which the South Bhowra area falls. In fact, JMS has two factions; one led by Kunti Singh – the former MLA of Jharia and wife of late Surydeo Singh – and the other is led by Bachcha Singh, the younger brother of Suryadeo Singh. The JMS factionalism is a direct outcome of the fallout between the two in the year 2005 when Kunti Singh contested the Jharia assembly elections, then represented by Bachcha Singh. Literally, the trade unionism in Jharia coalfield has been reduced to an extended family feud with these two groups trying to extract and grab as much of the resources as possible. Consequently, most of the contractors who get the outsourced projects enjoy patronage from either of these groups. This represents a web of a state–politician–mafia network which can at best be regarded as predatory. Violence is the method of reproducing this discourse and the bloody affairs between the two factions even recently led to one of many brutal murders.[11] One can easily notice the parochial method of working of both JMS factions and even though they always demand an increase in the statutory wages of the contract workers, it hardly reaches the targeted beneficiary. This "hybrid" labour discourse wherein the predatory CTUOs have emerged as the false messiahs of the contract workers needs to be dismantled. In fact, when compared to private enterprises, the presence of criminals has hampered the PSUs from generating profits despite endorsing policies like outsourcing.

Livelihood in the vortex of coalfields

While the study of collective action by contractual workers is itself a potential area of research in labour studies, the present chapter concentrates more on livelihood issues like working condition, wages and social sector benefits available to them. The livelihood options and patterns are very diverse and fascinating in the coalfield. Apart from the directly employed people in the coal mines, there are a whole range of activities which cater to the livelihood requirements. As outsourcing has emerged as a major means of employment in the coalfields, let us first discuss the laws meant to provide protection to the contractual workers and the major issues related to their working conditions. The legal protection to the contract workers comes from the Contract Labour (Regulation and Abolition) Act (CLA), 1970. However, one can easily notice the violation of rules and regulations as provisions for some of the most basic essentials like health centres and refreshment facilities are absent in the area. For instance, Chapter V of the CLA deals with welfare and the health of contract workers making provision for canteens, rest rooms and first-aid facilities. If these facilities are not arranged by the contractor, then the principal employer is supposed to provide them and deduct the cost involved from the payments to be made to the contractor. Even though the government has fixed a statutory wage of Rs 600/day to the contractual workers, the workers receive substantially lower wages between Rs 250 and 300/day. The workers are also made to work for 10–12 hours and under complete absence of social security. As there are widespread cases of underpayment in the South Bhowra Colliery, one needs to understand it against the existing statutory provisions. Clause 21 of the CLA, 1970 says that while the contractor is responsible to make payment of wages to contract labour, the principal employer shall nominate a representative to be present at the time of disbursement of wages and certify the same. The contractor should make sure that this nominee is present when the wages are being paid. In case the contractor makes a delayed or lower payment of wages, the principal employer shall pay the full and unpaid balance to the contract labourer and recover the amount from the contractor.

Clause 29 further makes the provision for records regarding the particulars, nature of work and rate of wages of the contract labour to be maintained by both the contractor and the principal employer. It is not very easy to procure the records from the contractor. However, even the mines management also looked reluctant to hand over the records and hesitatingly provided only verbal testimony for the underpayment made to the contract workers. This is largely true and has been cross-verified with the workers who claimed that they receive only up to 60 per cent of the statutory wages fixed for them. In fact, when compared to their counterparts employed in privately owned coal blocks, the contractual workers are exploited more. The private companies have been found to be paying wages somewhere around Rs 300–350/day in addition to certain allowances for medical and housing. This is one reason why the unemployed youths look first for a prospect in private companies and opt for the contractual employment only as a last resort. The government has, thus, not only disciplined the labour but has alienated them to an extent where they ask for more privatisation of coal mines rather than expansion of the existing PSUs. As far as unionisation among the contract workers is concerned, it is largely absent. However, the registered trade unions in the PSUs sometimes raise the issue of wages with an entirely different motive, which will be explained when we discuss the criminal, politics and trade union nexus in the next section.

Apart from the contractual workers, there remain a host of livelihood strategies in the coalfields of which we will discuss at least two which, though premised upon an *illegality*, are crucial in sustaining the local economy. The first of these relates to the transition in traditional occupations like farming while the second is associated with the mushrooming of coke plants in the neighbouring areas of West Bengal and Bihar. The first strategy can be theoretically understood as the "outside of capital"[12] of Kalyan Sanyal (2007) as these are the people who have not been absorbed by the capital after their land was acquired during "primitive accumulation".[13] Basically, these people are vegetable growers who cultivate the land once owned by them. A reference is being made here to the fertile tracts of land on the banks of river Damodar which once belonged to the local adivasi and non-adivasi people. However, it was acquired by the Damodar Valley Corporation (DVC) during the 1960s and 1970s. But the local people have continuously engaged in farming on their acquired land. Villages on the banks of Damodar river do not have access to proper irrigation facilities. Every year, after the river water retreats from the floodplains, the villagers grow vegetables. During the rainy season, the fisherfolk get a good yield of fish for at least three months. However, the annual flooding is welcomed by both the vegetable growers and the fishermen.

Another similar story from Andhra Pradesh wherein the land was acquired for the K. L. Rao Sagar multipurpose irrigation project establishes the utility of this "illegal" source of livelihood (*The Hindu*, 17 June 2017). As reported by Appaji Reddem, the farmers in the area continue growing vegetables on the acquired land which yields around Rs 50,000/ acre. Likewise, the coalfields are mainly surrounded by Santhal adivasi hamlets. Vast swathes of their land were acquired for the purpose of coal mining using the Coal Bearing Areas (Acquisition and Development) Act, 1957. Hence, one can notice a good number of Santhal adivasis employed in the coal mines mostly as manual labourers, owing to their poor education status. But not all the men who lost their land could be absorbed in the mines. Many of them could no longer grow paddy, which is the traditional crop, and hence shifted to growing vegetables. This shift proved to be beneficial as the people employed in the mines and living in the mining towns became their readymade customers. The sight of Santhal women selling vegetables in the weekly *haat* of these towns is very common. By selling vegetables, they can eke out their meager livelihood requirements.

This income, not to forget, is mostly supplemented with the earnings from the illegal sale of coal which the Santhal men procure through either illegal mining widely prevalent in the area or by stealing from the coal depots of the company. Apparently, who actually is stealing becomes a misnomer here. Kuntala Lahiri-Dutt has extensively analysed the issues of "illegal" or "unintended" mining by the communities and states that:

> The causes are buried under layers of complexities of outdated colonial laws of land acquisition and state-ownership of coal resources, lack of safeguards and protection of poor people, despicable social and environmental practices by formal mines, the disregard for social impacts by mining engineers and technologists, a continuity of licence "raj" in CIL, and the overall trend of informalisation of the economy. Illegal coal mines are an expression, locally, of unjust national mineral laws that fail to ask simple questions such as "who owns the mineral resources, since when and why", "who control their use", and "who is looting and under what circumstances"?
>
> (Kuntala Lahiri-Dutt, 2007, p. 58)

In calling the Santhal men thieves, one has to take a very legalistic view of the act of stealing despite the awareness that the government cannot absolve itself from the charges of theft as they failed to fulfil their promise of supplying these people with alternative sources of livelihood in lieu of their land. The government control over mineral resources is itself grounded in colonial history and has been contested by ethnic communities. There are success stories like that of Chitra[14] Collieries in the Deogarh sub-division of the Santhal Parganas under which 11 mines were operated by the Santhal adivasis and could have become a model for granting community mining rights. However, the government has shown a complete disregard for such innovations, obviously under pressure from larger forces. Nevertheless, the theft and smuggling of coal to the neighbouring sites of coke plants is widely prevalent in the coalfields. It is quite usual to see a boat loaded with coal crossing the Damodar river over to West Bengal where most of the coke plants are located. These coke plants thrive on both legal and illegal coal received from the coalfield and mostly belong to the local influential people. After the initial process of converting the coal into coke through their partial burning, it is sent to the brick kilns in West Bengal and other far off places like Varanasi, Allahabad, etc. This business is largely controlled by the coal mafia with the support of the local police and even the management of the PSUs. Apart from this, one can frequently see people carrying coal on overloaded bicycles to sell in the neighbouring areas. This is a more decentralised illegal practice wherein the poor people engage in illegal mining to procure coal or steal it from government depots. This coal is then sold in the towns to people running hotels, *dhabas* and other installations. Thus, the "political society" existing in the coalfields has to ensure its existence by negotiating their illegality of livelihood with the local state. This is easily done through bribing and sometimes asserting their political presence. However, the real impact of outsourcing has to be captured through the agony of the contractual labourers. Andrew Sanchez (2012, p. 443) rightly observes that the process of casualisation "closes the gap which ostensibly exists between the hegemonic consciousness of India's civil society and its 'political' counterparts".

Conclusion

The coalfields of Dhanbad began their journey as fiefdoms of the private miners during the colonial period. In post-Independence India, the quest of migrants from UP and Bihar to replace these miners through the use of organised violence dominated the scene. However, nationalisation of these mines led to the beginning of trade union politics. As explained above, different factions of society tried to establish their hegemony over these unions and also interacted differently with the local politics. In the post-liberalisation era, the decline of class-based politics further consolidated the grip of local mafias who had already made a stable place in the local politics. Policies like outsourcing have certainly changed the nature of politics in the coalfields, but the mafia-political elements found a new equation to work out their survival. They have now become the contractors for the outsourced works and suppliers of contractual workers. The labour politics has, thus, culminated in its collapse and has given rise to a political economy in which the state policies help the mafia elements to reproduce themselves. In fact, the coal mafia has become so influential that the tickets to Jharkhand assembly and the Lok Sabha seats are contingent upon the criminal background of the candidates.

The contractual workers live under pathetic working conditions and reproduce the local economy in a manner akin to that of the "outside of capital". But unlike the *political society*,[15] they lack the bargaining capacity to even ensure their welfare. The leitmotif of policies like outsourcing is to ensure the revival of the loss-making firms. However, the intricate linkage of mafia and trade unions hinders this profit-making mantra, proving outsourcing to be an uneconomical option. In fact, a better option to revive the PSUs would have been the outsourcing of managerial works to firms capable of supplying trained professionals. This approach would have saved the workers from the vagaries of uncertain jobs and liberated them from the clutches of corrupt and criminal trade unions. A committed army of middle and high ranking officers along with a determined state can actually help the revival of the PSUs which is crucial to ensure the energy requirements of a booming economy. This step will not only ensure a labour-friendly regime but also provide impetus to healthy unionism.

Notes

1 The National Commission for Employment in the Unorganised Sector noted that in the year 2004–2005, out of the total 54.9 million workers in the formal sector, around 23.1 million (42 per cent) are unorganised workers (2009, p. 13). The NCEUS report further argues that the growth in employment in the formal sector between 1999–2000 and 2004–2005 "has largely been of informal in nature i.e. without any job or social security". This is possible due to "greater flexibility of employment, unbundling employment, and an expansion in other types of outsourcing and contract services in both the secondary and tertiary sec tor of manufacturing" (2009, p. 14). Please see the link: http://sanhati.com/wp-content/uploads/2013/10/The_Challenge_of_Employment_in_India_Vol.I.pdf, accessed July 31, 2018.

2 Rajiv Kumar (1981) has identified at least six such reasons, i.e. failure of private operators to increase the output according to demand by the metallurgical industry; preponderance of small mines making scientific mining impossible; unwillingness of the private sector to increase the output of non-coking coal by 100 per cent; lack of huge investment capacity in private sector; malpractices on wages and other statutory dues to the workers; and finally, violation of mine safety laws.

3 The committee suggested a rise in coal production from 105 ml tonnes in 1973–1974 to 165 ml tonnes per annum by 1978–1979 to meet the future demands. The target was revised by the Draft Fifth Five Year Plan which kept it at 135 ml tonnes per annum. As there was already an excess capacity of 12.7 ml tonnes, it was suggested that an additional production of 17.6 ml tonnes by 1978–1979, adding annual capacity of 2.2 ml, would be sufficient.

4 Open cast mines are located where coal occurs close to the surface. The mining is done with the help of earth mover machines used to clear the overburden and carry the coal from blasted off seams. Usually, OCMs are around 100 metres in depth and not fit for extracting coal from deeper seams. It leaves a permanent mark on the mined areas and most of the coal producing countries have given up OCMs.

5 Scholarly works like Rudolph and Rudolph (1984) have highlighted the processes through which traditional practices interact with modern notions, thus transforming themselves in the process, but without completely acquiring a modern outlook. On the cultural aspects of change in the industrialised areas, Joseph (2004) takes note of the fluid cultural imagery created due to the interaction of many cultures under a new system of production. While it cannot be claimed that in such a hybrid social set-up, the traditional beliefs, systems, notions, and so on are completely uprooted, it certainly gives rise to new identities which can be framed in terms of a modified notion of tradition.

6 Rajesh Kalarivayil and Smitha S. Nair (2018, pp. 45–48) have identified four phases of trade unionism in India. The first phase spans between the 1950s and 1960s; the second phase extends from 1960s to 1979; the third phase covers 1980–1991 and the fourth phase extends between 1992 and 2000.

7 Ananya Bhattacharjee and Fred Azcarate (2006, pp. 67–68) argue that following the strategy adopted by NTUI, the garment sector workers, mostly women, in Bangalore were organised by Gopinath, the co-ordinator of Cividep which initiated the formation of a pre-Union Women Workers Forum that began addressing workers' issues outside the factory floor. Likewise, V. B. Cherian founded the Cochin Shipyard Employees Union and Milind Ranade formed the Contract Labourers Union in Mumbai Municipal Corporation and Nasik Municipal Corporation.

8 Vadim Volkov (2002) uses this term for the coercion and violence used by the violent actors in favour of the implementation of company policies and the smooth functioning of the enterprise. Such force is considered essential in disciplining the labour and its politics.

9 Andrew Sanchez (2016, p. 13) raises the issue of "episodic" and "systemic" corruption discourses. He argues that "while the former are popular ideas that model corruption chiefly in reference to localised and personalised acts of individual criminality, the latter is concerned with the mapping of functional relationships between different types of actors, and the relationship between corruption and seemingly legitimate enterprises".

10 Andrew Sanchez (2016) has regarded systemic corruption discourses as expressions of political consciousness which while articulating class struggle describe criminality as a socially constructed and embedded practice of modern capitalism.

11 Niraj Singh, the ex-deputy mayor of Dhanbad city, was allegedly murdered in a plot organised by his own cousin Sanjeev Singh who happens to be Jharia's MLA and the general secretary of Hind Mazdoor Sabha (HMS), the parent organisation of JMS. The political and economic discourse of coal money in Dhanbad is rife with such incidents.

12 Kalyan Sanyal (2007) refers to this category as comprising people who were dislocated from their traditional occupations but could not be absorbed within the new vocations owing to its lower capacity of employment. However, this category has found its way to sustain itself at the vicinity of capitalism by associating with the low-end jobs like street hawkers, domestic help and so on, thus making itself a part of the moral economy. This, Sanyal argues, plays a crucial role in sustaining the otherwise distressful capitalism.

13 It is a Marxian term used to denote the process of separation of the producers from the means of production.

14 For a detailed report on Chitra Colliery please see the article "Saharjori: Abandoned miners of Santhal Parganas" (1983). *Economic and Political Weekly*, *18*(40), 1690–1692. The article is available at the link: www.jstor.org/stable/pdf/4372550.pdf?refreqid=excelsior%3A8a1e85e9ec0c16d7b76ede2db0f30b26, accessed July 31, 2018.

15 A term used by Partha Chatterjee (2004) for the people who live in the slum areas largely as illegal migrants and occupiers of public space. These people, however, use their numbers for bargaining to get electricity, water, and other essential state-funded benefits from the elected representatives. Despite their illegality of space they have smartly sustained themselves by utilising the voting rights they hold.

References

Agarwala, R. (2006). From work to welfare: A new class movement in India. *Critical Asian Studies*, *38*(4), 419–444.

Bhattacharjee, A., & Azcarate, F. (2006). India's new unionism. *New Labor Forum*, *15*(3), 64–73.

Bhattacharjee, S. (2017). *India's coal story: From Damodar to Zambezi*. New Delhi: Sage Publications.

Chatterjee, P. (2004). *The politics of the governed: Reflections on political society in most of the world*. New York: Columbia University Press.

Gooptu, N. (2007). Economic liberalisation, work and democracy: Industrial decline and urban politics in Kolkata. *Economic and Political Weekly*, *42*(21), 1922–1933.

IEA. (2015). *World energy outlook*. Retrieved June 12, 2018, from www.iea.org/publications/freepublications/publication/WEO2015.pdf

Joseph, K. V. (2004). *Culture and industrial development: The Indian experience*. New Delhi: Anmol Publications.

Kalarivayil, R., & Nair, S. S. (2018). Unionisation in post-reform India: A review of trends and trajectories. In S. George & S. Sinha (Eds.), *Redefined labour spaces: Organising workers in post-liberalised India* (pp. 45–61). New Delhi: Routledge.

Khanna, A. A. (2013). *Governance in coal mining: Issues and challenges* (Working Paper No. 9). New Delhi: The Energy and Resources Institute (TERI).

Kumar, R. (1981). Nationalisation by default: The case of coal in India. *Economic and Political Weekly*, *16*(18), 824–830.

Lahiri-Dutt, K. (1999). State, market and the crisis in Raniganj coal belt. *Economic and Political Weekly*, *34*(41), 2952–2956.

Lahiri-Dutt, K. (2007). Illegal coal mining in eastern India: Rethinking legitimacy and limits of justice. *Economic and Political Weekly*, *42*(49), 57–66.

Ministry of Coal. (2012). *Report of the comptroller and auditor general of India on allocation of coal block and augmentation of coal production*. Retrieved July 31, 2018, from https://cag.gov.in/sites/default/files/audit_report_files/Union_Performance_Commercial_Allocation_Coal_Blocks_and_Production_Ministry_Coal_7_2012.pdf

Mohanty, M. (2009). *A note on trends in unionization in India* (Working Paper Series No. 641). Calcutta: Indian Institute of Management.

Munda, R. D., & Mullick, S. B. (2003). *The Jharkhand movement: Indigenous people's struggle for autonomy in India*. Copenhagen: International Work Group for Indigenous Affairs.

Papola, T. S., & Sharma, A. N. (2004, May). Labour: Down and out? *Seminar*, 537.

Rajalakshmi, T. K. (2016). A raw deal. *Frontline*, *33*(15), 36–38.

Reddem, A. (2017, June 17). Pulichintala's tenant farmers reap success as Krishna recedes. *The Hindu*. Retrieved from https://www.thehindu.com/news/national/andhra-pradesh/pulichintalas-tenant-farmers-reap-success-as-krishna-recedes/article19094760.ece (accessed 12 April 2018).

Roy, A. K. (2003). Disinvestment and outsourcing in coal. *Economic and Political Weekly*, *38*(49), 5153–5154.

Rudolph, L. I., & Rudolph, S. H. (1984). *The modernity of tradition: Political development in India*. Chicago: University of Chicago Press.

Sanchez, A. (2010). Capitalism, violence and the state: Crime, corruption and entrepreneurship in and Indian company town. *Journal of Legal Anthropology*, *1*(2), 165–188.

Sanchez, A. (2012). Questioning success: Dispossession and the criminal entrepreneur in urban India. *Critique of Anthropology*, *32*(4), 435–457.

Sanchez, A. (2016). *Criminal capital: Violence, corruption and class in industrial India*. New Delhi: Routledge.

Sanyal, K. (2007). *Rethinking capitalist development: Primitive accumulation, governmentality and post-colonial capitalism*. New Delhi: Routledge.

Sengupta, S. (1999). A regulatory body for the liberalised coal sector. *Economic and Political Weekly*, *34*(9), M25–M33.

Volkov, V. (2002). *Violent entrepreneurs: The use of force in the making of Russian capitalism*. London and Ithaca, NY: Cornell University Press.

12

FROM MOBILISATIONS TO MEDIATIONS

Shifting trajectories of the Latin Catholic Church's engagements with neoliberal development projects in Kerala[1]

Abdul Raoof C. K.

The role of religion and faith based organisations (FBOs) in development processes has been widely accepted although the modernisation project in post-Independence India tended to view religion within the strict framework of "secularism", that does not encourage religious interference in politics and governance. Undoubtedly, as in other parts of the world, it was expected that India would gradually develop into a "secular society", in which religion confines itself more to private spheres, in line with the postulations of secularisation theorists (Casanova, 2006, p. 9; Hovland, 2008, p. 17). However, as is well known, this demarcation in India has always been blurred and religious and faith based organisations played and continue to play active roles in developmental activities. Moreover, as scholars have argued, the relationship between the state and religion has emerged into complex forms of religio-political configurations (Menon, 2007). The space occupied by religious and faith based organisations in India is diverse and the involvement of religious organisations in democratic politics often crosses the boundaries of "secular spaces". Sometimes, they appear in public spaces as protesters of religiously ordained social issues (Puthenveed, 1985; Panikkar, 2009; Rao, 2016); they directly engage in political mobilisations, including elections wherein they play active roles from deciding the candidates to openly expressing support to certain political parties (Mahajan & Jodhka, 2012; Patel, 2010; Prakash, 2014) and candidates and take the role of active partners of the neoliberal state in implementing state policies and programmes (Hefferan, Adkins, & Occhipinti, 2009; Williams, Cloke, & Thomas, 2012; Occhipinti, 2015).

Several questions arise while examining these shifting approaches of religious organisations and FBOs with the state, its target constituencies and the larger society. What have been the trajectories of FBOs' developmental interventions? What determines their approaches on a particular socio-political and development issue? How do FBOs mediate between people and the state? Does the state use the moral authority and power of such organisations to pacify the discontent of people against neoliberal state projects? If so, how? This chapter takes the developmental interventions of the Latin Catholic Church (LCC), which has been playing an active role among fish workers in the coastal regions of southern Kerala, as a case in point to address some of these questions. We discuss the past interventions of LCC and examine their recent interventions in various social, political, environmental and livelihood issues in connection with the ongoing Vizhinjam International Seaport Project in Trivandrum, southern Kerala to understand the shifting trajectories of their development and engagements with the *neoliberal* projects of the state. We draw mainly from the field work conducted at Vizhinjam coastal area. The first section of the chapter delineates the various approaches of LCC to socio-political issues related to the fishermen community in Kerala from the 1970s. The second section examines how and why LCC intervened in a seaport development project at Vizhinjam, Kerala, and the responses from the affected people, the general public and the state authorities. The third section of the chapter examines how LCC manages the challenges raised by the state and the demands of the fishermen – which most of the times have been conflicting – simultaneously and the subsequent changes in its approach from confrontations and mobilisations to more passive forms of mediations and facilitations. This section also analyses how the state appropriates and manages the relationship between the FBOs and the state in the neoliberal era in order to effectively use them as part of its programmes.

Shifting trajectories of the developmental approaches of LCC

The present Trivandrum Latin Catholic Church was established by Pope Pious XI in 1937 by bifurcating the Kollam Diocese. The present revenue taluks of Chirayinkeezh, Thiruvananthapuram, Nedumangad, Neyyattinkara, and coastal parishes from Pallithura to Erayimmanthura come under its jurisdiction. Further, it expanded in 1952 by delinking coastal parishes from Kochi and adding to the Trivandrum Latin Catholic Diocese (Archdiocese, 2013). The Latin Catholic Church of Vizhinjam performs the ecclesiastical functions through the networks of parishes and Church hierarchies, and the non-ecclesiastical functions through its various FBO wings. Initially, the non-ecclesiastical policies and responses of LCC were criticised for being in favour of the elites in the society, because of their economic support to the Church (Reeves, Pokrant, & Mcguire, 1997). Later on, with their social service initiatives like establishing schools, higher education centres and hospitals, the focus of the non-faith activities of the Church had shifted to developmental philanthropy. By 1932, the Church owned 111 educational institutions, including primary schools, high schools and training collages. In later periods, LCC started adopting a rights-based approach rather than philanthropy-based social service, which was observed by many scholars in fishermen movements of the 1970s.

The Church has been involved in the fishermen's issues in south India since the colonial period in support of the Portuguese traders (Sundar, 2012). During the 1970s, LCC started mobilising fishermen of southern Kerala against the challenges raised by the modernisation project in the fishing sector in 1952.[2] The modernisation policies encouraged intensive fishing methods like deep sea fishing, test fishing and joint ventures (Reeves et al., 1997). The new methods of fishing with mechanised trawlers and fishing vessels helped in increasing production; nonetheless, it adversely affected the traditional fishermen in terms of reducing catchment, availability of fish and marine resources. More often, the traditional fishermen could not afford mechanised boats, which were mostly owned by middlemen or the new working class, and the traditional skills and knowledge were replaced by the mechanised system, which also alienated the traditional fishermen (Meynen, 1989; Sinha, 2012; Sundar, 2012; Reeves et al., 1997). Besides, the incidents of death and property/livelihood loss of traditional fishermen due to collision of traditional boats with mechanised boats during shore-line fishing triggered more tensions. The policy of modernisation by mechanisation to increase productivity led to strong protests in other sectors like agriculture as well. Political parties like the Communist Party of India (Marxist), CPI(M), organised farmers and industrial workers under their trade unions; however, such arrangements and organisations were absent for the fishermen. LCC, with the active involvement of some priests, took up the matter of organising the fishermen community and succeeded in establishing the first fishermen union, called Kerala Latin Catholic fishermen forum (KLCFF) in 1977 at Alleppey. The KLCFF submitted a memorandum to the chief minister of Kerala demanding restriction of the mechanised boats within 5 kilometres from the shoreline, ban on trawling during the monsoon season[3] and curbing of inland water pollution (Roshan, 2011). In 1978, they joined hands with similar groups from Tamil Nadu and Goa, which were also historically strongly influenced by Latin Catholic missionaries, and formed the National Fishermen's Forum (NFF). The organisation set up district-level units like the Trivandrum District Fishermen Union, which assisted in all the ground-level mobilisation of KLCFF. Later, they joined the World Forum of Fish-harvesters and Fishworkers in 1997. In 1980, KLCFF changed its name to Akhila Kerala Swathanthra Matsya Tozhilali Federation (AKSMTF) as suggested by the leaders of the organisation, to attract the Hindu and Muslim fishermen. The organisation adopted a secular outfit by affixing the term "Swathanthra" – which means independent in Malayalam – to show their "distance" from all other religious and political identities (Roshan, 2011). But the leaders were confused whether a priest or layperson should take charge of the organisation. Even though Fr Parisavila was appointed as the head of the AKSMTF, the dissent among the leaders divided the organisation into two sections. One section identified as "radicals" believed in a more liberal form of leadership and conducting strong agitations for the issues of fishermen. The other section was known as "loyalist" (according to radicals), and they believed in remaining under the leadership of the Church and keeping the political alliance with the Congress-I. The loyalists dominated the leadership of the AKSMTF and influenced the Church hierarchy to control the radical priests like Fr Kocherry from mobilising fishermen of their parish churches (Reeves et al., 1997). The influence of the liberation theology and working class mobilisation of left-wing political parties influenced these priests and nuns involved in organising fishermen. The common causes of fishermen made them think beyond the interests of religion and political organisations (Puthenveed, 1985).

The relationship between LCC and the state, as it is clear from the events that led to the formation of NFF, were confrontational in nature. For instance, on the neglect of the memorandum submitted in 1978, KLCFF protested through a two-month long hunger strike against the chief minister. They also took out a rally from Alappuzha to Trivandrum to mark their protests against the death of a fisherman due to the collision of his boat with a trawler. Similarly, they strongly demanded the enactment of the Kerala Marine Fisheries Regulation Act,[4] ban on monsoon trawling, and establishment of welfare and housing schemes for fishermen. As a result of continued agitations, in 1981, the Kerala government announced a ban on monsoon trawling. The agitation continued against the relaxation, which was allowed later at *Neendakkara*,[5] and forced the state to appoint a committee to look into the matter.[6] The committee submitted its report; however, its recommendations got implemented only after another agitation. AKSMTF invoked different modes of protest, like picketing ministers' residence, indefinite fasts, and surrounding the district collectorate, rallies, demonstrations, torch processions, door-to-door campaigns, mass sit-in in front of collectorates and picketing of national highways. They called off the strikes when the chief minister agreed to ban night trawling and allocate 80 million rupees for fishermen welfare programmes. In 1985, AKSTMF resumed the agitations in the form of the hunger strike of Fr Thomas Kocherry,[7] filling-the-jail campaign, marches to district capitals, and surrounding 125 fishing harbours with country boats. This succeeded in extracting a temporary ban order on monsoon trawling and appointment of a committee specifically to resolve the matter. The committee submitted the report in 1989 and recommended the ban.[8] The AKSMTF further intensified the agitation and finally succeeded in getting a cabinet order which put a permanent ban on trawling between 20 July and 31 August every year.

The rights-based approach and the confrontational engagements of the Church with the state machinery continued all through in their work with the livelihood issues of traditional fishermen in the region. For instance, the Church along with KSMTF and NFF continued raising the issues of fishermen and staged agitations against coastal pollution, and the Kudamkulam nuclear plant.[9] At one point of time, the intervention even took violent forms. Eight fishermen were killed in police firing. The post-1990 fishermen's movements became more violent than those in the 1980s, which was the period of agitations while the 1970s were the formative years of the fishermen's movement with various organisations consolidating themselves (Reeves et al., 1997). The other trade unions like CITU[10] and INTUC[11] were critical of such interventions of NFF and KSMTF supported by the Church (Radhakrishnan, 1997). The Church not only mobilised the fishermen for various issues, but also facilitated fishermen to cope with the new system by providing training to handle the mechanised boats, which were accessible only to a handful of fishermen (Sundar, 2012).

With such an active and direct confrontational approach with the state in the everyday survival issues of the traditional fishing community, the LCC developed an *influential* relationship with the fishermen and other religious communities, including Muslims and Hindus. To quote Father Matthias,[12] "our involvements in social issues are part of our moral obligation for supporting the deprived communities, and this commitment may be one of the factors that made us more trustworthy than political leaders".

These relations helped in bringing them together for discussing regional issues related to livelihood and other developmental needs. Although such a collaborative approach of LCC with other socio-religious groups of Vizhinjam resonates with Fulton and Wood's proposition of *Progressive Churches*, which refers to viewing social issues beyond the narrow-minded community-centric understanding (2017), LCC went even further in its engagements. They adopted a trade union mode – negotiate for issues of fishermen, discuss major changes to the workplace, provide members with legal and education facilities – to actively engage with the state for the rights of its serving constituency (read fishing community). While doing so, the Church also insulated itself from becoming openly "too political" by not directly spearheading these movements, but by actively engaging through feeder organisations, like Coastal Watch, newly formed councils, Centre for Fisheries Studies (CFS), Young Men's Christian Association (YMCA) and so on. In other words, the Church utilised the social spaces of FBOs, which helped the Church in gaining public acceptance across socio-religious and political groups. Such platforms created by LCC have critically looked at the policies of the state pertaining to them as well as the politicians who do not attend to their concerns and might have compromised with vested interest groups. This was an opportunity for LCC to expand their role from a conventional FBO to a more politically active group, which could mobilise people for common concerns by transcending their faith boundaries. Eventually, the interventions of LCC covered many other common issues, ranging from the demand to construct fishing harbours and the resettlement issues of 1990s, the demand to arrest the culprits and assuring justice to the victims of communal conflicts of 1990s, mediating/organising peace talks, to addressing issues such as the outbreak of diarrhoea and cholera cases in 1994. On health-related issues, their strategy was inclined towards a rights-based framework to engage with the state through various channels.

The port project and the responses of Latin Catholic Church: the moments of change

The proposed Vizhinjam International Seaport Project (VISP),[13] which lies adjacent to Vizhinjam fishing harbour, came to mainstream developmental discourses through the campaigns of organisations like LCC, Vi-MAC, Janapaksham and political parties. Contrary to their commitments to the livelihood and environmental struggles, the Church took a position in favour of the port project, subscribing to the carefully constructed dominant view that it would lead to employment generation for the fishermen and the overall development of the area. Interestingly, the Church used its organisational capabilities, influence and even moral authority on the Church members to create opinion in support of the project initially. For instance, the Church had mobilised fishermen in huge numbers to raise the demand, through its public platforms, like Coastal Watch, CFS, Trivandrum Matsyathozhilai Federation (TMF),[14] YMCA, and through newly formed platforms like Theeradesha Mekhala Samyuktha Samithi, Vizhinjam Action Council, etc. The Church also projected publicly its "pro-development approach" by drawing attention to cooperation with projects like the Trivandrum International Airport, Indian Space Research Organisation (ISRO) and Vizhinjam seaport.[15]

Consequently, the residents of Vizhinjam including the Church-supported fishermen organisations took up the demand for the port project after commencement of the fishing harbour project in 1991. This was based on a widespread perception developed among the local residents by comparing the condition before and after the fishing harbour. They believed or were made to believe that the construction of a transshipment port at Vizhinjam would bring many advantages in terms of employment creation and infrastructural development. Such narratives were heard from many quarters of Vizhinjam. For instance, in our interview, a representative of the business community, Ms Prameela Rajan[16] noted that "the port project is going to be a great step to development of Vizhinjam as well as Kerala, our next generation is going to be benefitted as the fishing harbour helped us in overcoming poverty". Some of the discourses developed during this period over the seaport project were: the Vizhinjam transshipment port could give competition to international port cities, like Singapore, Dubai and Colombo and the employment and the economic income in terms of various taxes and charges will boost the GDP as well as livelihood of the locals. On the same lines, the project was introduced as a "Green port project" by which the locals understood it to be an eco-friendly project while "Green port project" is merely a technical term referred to the projects which need no remodelling or demolishing of existing infrastructure.

In 2005, the Church mobilised the masses by conducting awareness campaigns about the benefits of the project for the local residents and the development of the region among the public as well as among the fishermen community which would be replaced due to the port project. No political parties or state apparatus questioned these public interventions of a religious body; rather, they welcomed the patronage of the Church. The Church and its feeder organisations had been instrumental in developing favourable public opinion for the port in the initial years. For instance, Alias John, the president of Vizhinjam Motherport Action Council (ViMAC),[17] recollected the support extended by the Church authorities for the activities of ViMAC, an organisation that actively campaigns for the construction of the port project. He notes that the ViMAC was founded in the presence of the head priest, Rev. Dr. Soosa Pakiam, and representatives of all other socio-religious groups at Trivandrum Bishop House. The Archdiocese of Trivandrum gave the correspondence address of ViMAC. The programmes of ViMAC were conducted at schools run by the Church, or, at Church premises with the participation of priests. These campaigns have been successful in constructing the image of the port project as a "dream project" for the larger Kerala society rather than a regional developmental question. It even became one of the key topics of the election campaign during the legislative assembly election of 2006. On November 11, 2006, LCC publicly criticised the newly elected government for cheating the fishermen by not considering the demand for the seaport after pooling votes with the offer of the materialisation of the port project (Deshabhimani, 2006). During this period, many other organisations also raised a demand for the project except the Kerala Swanthantara Malsya Thozhilali Federation (KSMTF)[18] which raised concerns over the possible livelihood loss of traditional fishermen due to the project.

The state government started responding to the demands of these agitations, after which VISL[19] developed the design, notified international tender, appointed the agencies to conduct a feasibility and socio-environmental impact study and started identifying required land. Such rapid procedural development and greater support was unexpected from the LDF-led[20] government. Many were of the view that the initiation of the project was an "electoral gimmick", as it got delayed earlier due to various reasons. In addition, an international transshipment port at Kochi and Colachal[21] was under consideration by the central government. However, the state government's decision to roll out the project provoked many environmentalists and others including some of the Church-backed fishermen organisations. Gradually, these organisations started influencing the Church to rethink its position in favour of the project and inform the fishermen about the possible livelihood loss and environmental damage due to the project. Even though the organisations like Trivandrum TMF and Swathanthra Malsyathozhilali Federation (SMTF) raised their anxieties, the Church encouraged the people to cooperate with the project. For instance, Thomas Mar Makarios[22] asked people of Vizhinjam to stand together for regional development (indicating support to the construction of the port project) during their annual conference (*Malayala Manorama*, 23 February 2006). Since these mobilisations were based on developmental needs of a region, LCC was not criticised by any political party for their public interventions; rather, they considered it as a potential partner in developmental endeavours of the state. Scholars observe that FBOs have been incorporated by the neoliberal states as an active partner in welfare provisions (Williams et al., 2012). Similar interventions of FBOs were considered as a positive shift in the philanthropic approach of LCC towards developmental approaches.

The support of LCC for the project, however, was questioned by a faction inside the Church, since the concerns of livelihood loss and displacement started mounting from the affected people. For instance, in 2008, a sub-diocese of LCC submitted a memorandum to the state demanding that the seaport project should not displace the people from densely populated locations like Vizhinjam. The memorandum sought relocation of the project from Vizhinjam, which was also raised by Coastal Watch, another feeder organisation of LCC, later. It urged LCC to cooperate with new forums like *Janakeeya Kootaima Samrakshana Samithi* (people's collective for protection council), Action Council, *Bhoo Samrakshana Janakeeya Kootaima* (people's council for protecting land), which raised the demands of livelihood and environment protection. The dissent to the port project mounted among environmental activists and the local residents of adjacent villages dominated by Latin Catholics.

As a result, a fraction of LCC managed to bring together a mass from the fishermen community in 2013, and filed an appeal at the National Green Tribunal against VISL and the state government indicating the violation of coastal regulation zone norms[23] with the establishment of the port. They questioned the economic viability of the project by drawing from the feasibility study reports submitted to VISL by the International Financial Commission (IFC) as well as the environmental and social impact study of L&T.[24] The Vicar General of Archdiocese, Monsignor Eugene Pereira, publicly expressed the view that the Church was worried that the proposed seaport project would affect the livelihoods of fishermen (*Deccan Herald*, 12 July 2013). These developments forced LCC not to indiscriminately support the project and hold an open discussion with the state on some of the concerns raised by the affected people while other socio-political and cultural organisations including political parties and the Vizhinjam diocese hesitated to take up these demands.

By this time, the protest against the project reached its peak with the entry of other civil society and religious organisations. Organisations including Muslim Jama'at, caste-based organisations like Dheevara Sabha and Nadar Association, for instance, started mobilising the affected people against the port project. People from adjacent villages, like Panathura, Kovalam, and Punthura were also mobilised along with environmental activists, like Sahyaparvatha Samrakshana Samithi. The LCC-backed organisations, like the Coast Watch (CW), Trivandrum Matsyathozilali Federation (TMF), Swathanthra Matsya Thozilali Federation (KSMTF), Trivandrum Social Service Society (TSS), and Centre for Fisheries Studies (CFS) also joined these campaigns. All these forced LCC to revisit its position on the port project. Eventually, LCC began to withdraw from openly supporting the port project and started accommodating the dissenting voices of its feeder organisations and internal factions.

Pacifying discontent: LCC mediating between the state and the project affected

Although the dissenting voices against the project within LCC have gradually gained momentum over a period of time, it is incorrect to say that the Church as a whole accepted the changed position of its internal faction opposing the port project. The deviation from the earlier position of LCC created further conflicts in the Church due to the mounting pressure from inside and outside. For instance, although the LCC had taken a decision to pinpoint the problems of land acquisition, rehabilitation and environmental damage due to the project, the parish church of Vizhinjam had taken a different stand on the port project. The secretary of the Vizhinjam parish church in our interview noted that some of the fishermen believed that the project would not harm their livelihood or living environment. The parish church did not cooperate with the archdiocese and denied reading the pastoral letter demanding withdrawal of the project in its then form. Instead of supporting the archdiocese, the Vizhinjam parish church protested in front of archdiocese headquarters at Trivandrum in 2014. Similarly, the state government and the leaders of socio-political organisations began to criticise the stand of LCC and started to stay away from the public interventions. On 21 July 2015, LCC announced that they were proceeding with a series of protests against port projects, since the government had not conducted any proper social impact study (Malayala Manorama, 2015). The critiques of LCC's stand against the port project invoked the role of Latin Church in Koodamkulam movements of 2011 to portray them as an anti-development organisation. By 2014, organisations like ViMAC, Bharatiya Janata Party

(BJP), and other politicians accused the Church and claimed that the LCC is anti-development, and an agent of foreign forces being funded by international lobbies, and supporters of the resort mafia of Vizhinjam. LCC realised that such outright criticism of the project was defaming the Church in the eyes of the public, and it might delegitimise their interventions as well. The Church defended the allegation in a press meeting and the Archbishop, Dr Soosa Pakiam, made their compromising stand on the project clear by stating that:

> We are not supporting or opposing the international seaport project but our concern is the displaced and the affected fishermen should get equal compensations and proper resettlement packages.
>
> (*Manorama Online*, 31 July 2015)

Subsequently, LCC started diverting itself from the demand for the withdrawal of the port project towards demanding more resettlement and welfare schemes from the state, even though they did not withdraw the court cases. On 1 August 2015, LCC issued a pastoral letter challenging the port project and on 10 August, 2015, the Vicar General of the Diocese, Eugene Perera, urged the leaders of the fishermen community to mobilise the community (*Malayala Manorama*, 2015). On 13 August, 2015, the Kerala Catholic Methran Sabha also asked the state government to ensure resettlement packages before starting the port project. The SC/ST Service Society further demanded a proper study of port-affected people and resettlement packages on 22 August 2015. In 2015, the chief minister of Kerala conducted a series of meetings with the protesting organisations, including LCC, in which he met the Bishop and the caste leaders separately. The LCC gave tacit approval for the project based on certain conditions of ensuring fair compensation and resettlement packages. The withdrawal of LCC from the forefront of protest was reflected in Church-backed dissents. For instance, the Punthura parish, which was worried about their village being wiped out by accretion and the mini fishing harbour which was overshadowed by the Vizhinjam fishing harbour, shifted their demand to the rehabilitation package. However, the environmental activists, and the local residents of Punthura village, still oppose the port project. The "careful" silence of the Church, after being assured of the resettlement packages, was criticised by many environmental activists of the "No Vizhinjam" campaign,[25] as a compromise with the state.

The construction of the project finally started on 5 December 2015 by Adani Ports Private Ltd. and the people halted their protests for the time being, believing the packages offered to them. However, as reported by the respondents, the rehabilitation packages got delayed, many were removed from the list of who all the project affected, and sometimes even the entry to sea was denied to traditional fishermen. During our visit to *Matsyafed*[26] as part of the fieldwork, the people who came to avail their welfare schemes expressed their concerns. The disappointment of the people seemed to have mounted over time, and many noted that "the project is not going to employ them as it was expected and they are going to lose their access to sea as well as land". We argue that the evolved neutral stand of LCC and the state support to the project made the dissenting voices of fishermen organisations and environmental activists weaker. This was the moment of shift in the developmental approaches of LCC. They became more entrenched in the roles of a facilitator and mediator of a neoliberal state facilitated business project. It was a marked shift from its earlier rights-based and confrontational interventions with the state, yielding to the criticism on the legitimacy of LCC.

Changing roles of FBOs in neoliberal states: a discussion

The changes in approaches of LCC to the port project raise several issues with regard to the changing nature of the relationship between religion, FBO and the state. First, what are the factors that influenced the state and the socio-political organisation to criticise the social interventions of LCC? Second, what are the conditions that allowed the state to incorporate faith organisations to state services and public interventions? Lastly, what influenced the LCC to move from a critical opposition stand to demanding certain welfare schemes?

The epistemic base of such shifts in strategies and approaches of FBOs in general was a common trend across the world with the developments, as can be seen in the rise of US Christian rights during the 1980s, structural adjustment and the rise of NGOs in the 1980s, rise of political Islam, decline of communism and the rise of identity politics, emergence of Transnational Civil Society (TCS), and a growing multi-cultural, faith community due to immigration (Clarke, 2008). The shift in the LCC's approaches to developmental activities began when they laid the foundation for the trade unions among fishermen of southern Kerala which was followed by other political parties also (Sinha, 2012; Reeves et al., 1997; Sunder, 2017). These variegated roles of the Church make it difficult to categorise LCC as a mere religious institution, or, social service organisation, as Tomalin (2013) argues that the demarcation between FBO and religion is more or less confusing. Although the Church was one of the active agents in various social services, like establishing educational institutions, hospitals, and organising the fishermen community for their issues, it was generally believed that faith has no role in the new secular political system as an influential factor in determining the state policies and life of the citizen in the modern political system (Hefner, 1998, Clarke, Jennings, & Shaw, 2007; Tomalin, 2013). Hence, considering LCC as one of the FBOs, it tends to reformulate and be responsive to the changing socio-political scenario and to the needs of its target group within an acceptable outfit. In short, the changes in the nature of LCC's public interventions since the formation show its transition in tune with the nature of the state for redefined legitimacies.

Scholars observe that the nature of the relationship between FBOs and the state has been changing according to socio-political conditions existing in society (Jodhka & Bora, 2012), which is evident from the response of LCC to the seaport project in the last 26 years. The materialisation of the seaport project, which was delayed since 1991 due to various reasons including non-cooperation of the central government, was considered as an achievement of the UDF-led government in Kerala. The bridging factor which helped the state government to roll out the project was the re-introduction of the Sagarmala[27] project by the union government in 2014. The BJP-led government showed a combination of right wing–neoliberal economic policies in all sectors (Varghese, 2014) and it sought to boost up the blue economy by intensive industrialisation of the coastal area (Budhraja, 2017). This policy led to the rebirth of the Vizhinjam port project, as the Sagarmala project included renewing of existing ports and establishing large ports all over the coastal regions. This policy shift was reflected in the Vizhinjam seaport projects also, as the project which was stuck in redtapism got technical clearance from the ministry as soon as it changed from a public sector investment to the Public Private Partnership (PPP) model. The tender notification issued by VISL got good responses from various port running companies; however, only three companies submitted the bid application. VISL processed the discussion with them without going for a re-notification, and blamed the protesting nature of the state as a reason for the step back by companies. The court cases which were under the consideration of the NGT and the Supreme Court had also raised worries among VISL and the companies that participated in the bidding. The criticism against the LCC and related organisations was raised during this period of time. Both the central government and the state government intended to quiet them about the cons of the port project. Scholars had observed the nature of the neoliberal state in incorporating third sector organisations[28] and strategies of controlling their resistance to the state. The new political rationality of capitalism has carefully incorporated FBOs into state welfare services, believing that the ethically flavoured activities are able to shape, and be shaped by, the rationalities of the state (Williams et al., 2012). The incorporation of the FBOs into the state system has to be understood as the influence of the changes that happened at larger economic and political policies. Hence, the welfare services of third sector organisations had contributed to bridging the gap created by the withdrawal of the state (Williams et al., 2012; Occhipinti, 2015). Though the subjugation of third sector organisations to the neoliberal rationalities was expected, the socio-political interventions of LCC were influenced by *liberation theology*, which motivated them to stage mobilisations for the rights of fishermen since the 1970s. In other words, even though FBOs had taken a participatory role in the state service delivery system, they also had tried to contest with the state policies. However, the participation of FBOs as agents of state in India and other third world countries had also been discussed as part of "religion and development" discourses (Bano, 2007; Mahajan & Jodhka, 2012; Tomalin, 2013), and their involvement in political mobilisation as "progressive activism"[29] (Fulton & Wood, 2017; Fuist, 2017).

Studies reveal a growing relationship between religion and the political state in the neoliberal system as a strategy of appropriating moral authority for neoliberal rationality. The neoliberal system has institutionalised a particular configuration of secular and religious spaces, wherein the neoliberal technologies govern both individuals and groups according to certain political logic which need not to be secular or religious but fostering neoliberalisation (Williams et al., 2012). For instance, it is evident from the recent behaviour of the union government, where it used state technologies of governance to control all third sector organisations like non-governmental organisations (NGO) and FBOs challenging state policies. In 2014, the union government amended the Foreign Contribution Regulation Act (FCRA), which restricted many NGOs from receiving foreign aid. Within two years of the amendment, more than 20,000 NGOs' FCRA licences were cancelled. The Intelligence Bureau (IB) labelled them as agents of foreign governments, and anti-national. Moreover, IB reports sent to the Prime Minister's Office and other government agencies claimed that the activities of these organisations, like agitations against nuclear and coal-fired power plants, environmental protection and human rights violation across the country, have negatively impacted GDP growth by 2–3 per cent.[30] According to the open letter written to the Prime Minister by a host of civil society organisations, the reason for the new amendment is given as:

> Our work for the poor and the marginalised might involve questioning and protest decisions taken by the government, and this is our right. The government may not agree with what our policies are, from opposing nuclear power plants to campaigning to the right to food. Yet we expect the government to protect our democratic right to protest without being targeted as anti-national.
>
> (Chandhoke, 2017)

These newly created political conditions put LCC in a dilemma of continuing with the protest and opting for a safe (legitimate) way forward. Although LCC has managed to defend the allegations of being anti-national and an agent of western countries, it took a step backward based on the state offered welfare packages and NGT verdicts from its earlier rights-based confrontational position. We note that the silence of LCC is not a conscious compromise with neoliberal state policies, but a state imposed sanction through various direct and indirect measures. Though the exercise of power through the governmental administrative apparatus is more often considered as successful in meeting the objectives, the resistance of FBOs, nonetheless, led to revisions in state policies. The demands raised by LCC for more inclusive compensation packages, a mini fishing harbour at the adjacent village, concerns of accretion at southern coasts, etc. were met during the implementation of the project.

Conclusions

The Latin Catholic Church has adopted an indistinct stand on the construction of the seaport project while it has firmly and constantly taken a pro-environmental approach and stood by the livelihood rights of affected people on other occasions. We argue that in this particular case, the position of the Church evolved by yielding to the pressure of the state and the capital. For instance, during the 1990s, the Church strongly criticised and opposed the state government's decision to go ahead with the proposed thermal plant cum seaport project. The Church succeeded in mobilising the public by highlighting the hazardous effects of pollution on natural resources, more specifically on the availability of fish and drinking water, which forced the state to cancel the project. The Church supported the demand for the seaport project in the initial phase, but later on changed its position twice – first to oppose it due to the pressure from the non-loyalist faction within and next to reconcile with the project. Further, the strong resistance from the Church-backed organisations also became a challenge in materialising the project. However, in this case, the Church shifted its position, and started demanding welfare schemes for the affected instead of completely opposing the project at a later stage. The dwindling positions of LCC reflect new power relations developed between the state and FBOs in Kerala. Although the FBOs tend to focus on their core philosophy, the neoliberal state systematically incorporates them as agents of state in various service delivery systems. While doing so, the state also manages the discontent of the people against its projects and programmes, using the same platforms of FBOs, their influence and moral authority over the people.

Notes

1 This chapter is developed from the data collected as part of my doctoral fieldwork and I am thankful to Dr Sobin George, my PhD supervisor for his comments on the chapter.
2 The process was accelerated by rapid mechanisation of fishing through the Indo-Norwegian Project (INP) to increase productivity and employment.
3 The main objective of this "seasonal" ban on fishing is to protect fish wealth while commercially important species breed.

4 To keep big vessels 22 kilometres away from the shore lines and earmark the boundary for traditional fishermen. It was enacted in all the states except Gujarat. For details, see Radhakrishnan (1997).
5 Neendakara is a famous fishing port, located 9 km north of Kollam district in Kerala, with the capacity to contain over 500 fishing boats at a time. The headquarters of the Indo-Norwegian Fisheries Community project, established in 1953, was based in Neendakara until 1961.
6 During the 1980s, three committees were appointed by the state, but only the Balakrishna Nair Committee of 1989 considered the demands of fishermen. For details, see Reeves et al. (1997).
7 He was a Catholic priest and active member of Trivandrum District Fishermen's Union and moved to NFF due to some internal conflicts in AKSMTF in 1983 and became the president between 1982 and 1996.
8 Three committees were appointed by the state of Kerala to study the issues concerned with fishing; the Babu Paul Committee in 1981–1982, the AG Kalwar Committee in 1984–1985 and the Prof Balakrishnan Nair Committee in 1989.
9 Koodankulam nuclear power plant is the largest nuclear power station in India, situated in Koodankulam in Tirunelveli district of the southern Indian state of Tamil Nadu. The project was opposed by the local fishermen.
10 Centre of Indian Trade Unions (CITU), a trade union affiliated to the Communist Party of India (Marxist), and INTUC are among the active trade unions in Kerala.
11 Indian National Trade Union Congress (INTUC) was founded on 3 May 1947 as a trade union wing of the Indian National Congress, and is affiliated with the International Trade Union Confederation.
12 Former director of Centre for Fisheries Studies, Trivandrum.
13 Vizhinjam international seaport project was proposed after the establishment of the Vizhinjam fishing harbour in 1991. Even though the project was proposed much earlier in the 1990s based on the natural features of Vizhinjam and developmental needs of Trivandrum, the construction got delayed due to different reasons and the project was still under construction in 2018.
14 Trivandrum Malyathozhilali Federation is the fishermen wing of Trivandrum Social Service Society working under LCC.
15 The Church defended the criticism of "being an anti-development agent" by indicating that they have already cooperated with the land acquisition for airport construction and the establishment of ISRO in Trivandrum. When a demand for land was raised by the state, the Church has cooperated by giving acres of its own land even though the projects didn't help the Church in any way.
16 She is the president of the women's wing, Vizhinjam unit of Viyapari Vyavasayi Ekopana Samithi (VVES). VVES is the trade organisation formed in 1980 to safeguard the interests of traders, businessmen and industrialists of Kerala.
17 Vizhinjam Mother Port Action Council (ViMAC) is an organisation established in 1994 as a conglomeration of several bodies including various NGOs and religious groups demanding clearance for the Vizhinjam port and its fast implementation.
18 The Kerala Swatantra Matsya Thozhilali Federation (KSMTF, or the Kerala Independent Fish Workers Federation) is a trade union of small-scale artisanal fishers of the state of Kerala in south India under the influence of LCC.
19 Vizhinjam International Seaport Limited was established in 2004 as a special purpose vehicle/nodal agency to materialise the seaport project.
20 Left Democratic Front (LDF) is one of the two major political coalitions in Kerala, led by the Communist Party of India (Marxist) (CPI-(M)). Besides CPI (M),

the alliance includes Communist Party of India (CPI) and other smaller parties like Janata Dal (Secular), Nationalist Congress Party, Communist Marxist Party (Aravindakshan), Congress (Secular) and Kerala Congress (B).

21 This is a port on the Malabar coast, 20 km north-west of Kanyakumari, the southernmost tip of peninsular India. The port also has the natural advantages of Vizhinjam. Tthe Colachal port has been raising a challenge to the viability of Vizhinjam port since 2006.

22 He is the founder of the diocese of the Malankara Orthodox Church in the United States and Canada.

23 Construction activities in the notified coastal zone are highly restricted in CRZ regulations of 1991 which were amended in 2011 by removing the term "areas of outstanding natural beauty" in which Vizhinjam falls. This was raised to the principal bench of the NGT by Advocate Raj Panjwani. This argument was accepted by the court and it said that the NGT can revisit the CRZ 2011 amendments if it is affecting such areas.

24 L&T Infrastructure Engineering Ltd is one of the engineering consulting firms offering technical services in transport infrastructure which includes DPR & master planning of ports, tender and detailed engineering, environmental studies and development of greenfield ports.

25 The campaign was organised by Theeradesha-Sahyaparvatha Samrakshana Samithi at Trivandrum Press Club.

26 Matsyafed is the Kerala state Co-operative Federation for Fisheries Development Ltd, the apex federation of 654 primary fishermen co-operative societies spread over 10 districts of Kerala.

27 Sagarmala project is a US$120 billion investment initiative of the Government of India for setting up more than six mega ports, modernisation of existing ports, development of 14+ Coastal Economic Zones and 29+ Coastal Economic Units. It also includes the development of mines, industrial corridors, rail, road and airport linkages with these water ports to increase export revenue and generate direct and indirect jobs.

28 "Third sector organisations" are organisations that are neither public sector nor private sector, including voluntary and community organisations, social enterprises and co-operatives.

29 As religion is usually tied to conservatism, progressive activism as the important way in which religion and politics intersect is being overlooked. It tries to reveal that progressive religious activists are a driving force in American public life, involved in almost every political issue or area of public concern.

30 See the *Times of India* report on 12 June 2014, *Firstpost* on 30 December 2016, *The Guardian* on 7 September 2016, *The Hindu* on 20 July 2017 and 2 February 2018.

References

Archdiocese. (2013). *Socio-economic and pastoral survey 2011: A report and analysis*. Trivandrum: Latin Archdiocese of Trivandrum.

Bano, M., & Nair, P. (2007). *Faith-based organisations in south Asia: Historical evolution, current status and nature of interaction with the state*. Birmingham: International Development Department, University of Birmingham.

Budhraja, H. M. (2017). *Ocean-based blue economy: An insight into the SAGAR as the last growth frontier [Blogpost]*. Retrieved January 21, 2018, from NITI Aayog. www.niti.gov.in/content/ocean-based-blue-economy-insight-sagar-last-growth-frontier#

Casanova, J. (2006). Rethinking secularisation: A global comparative perspective. *The Hedgehog Review*, *8*(1–2), 7–23.

Chandhoke, N. (2017, July 20). The big squeeze on civil society: On the right to freedom of expression. *The Hindu*. Retrieved from https://www.thehindu.com/opinion/lead/the-big-squeeze-on-civil-society/article19309649.ece (accessed 16 March 2018).

Church Worried over Vizhinjam EIA. (2013, July 12). *Deccan Herald*, Thiruvandrum.

Clarke, G. (2008). Faith-based organisations and international development: An overview. In *Development, civil society and faith-based organisations* (pp. 17–45). London: Palgrave Macmillan.

Clarke, G., Jennings, M., & Shaw, T. (Eds.). (2007). *Development, civil society and faith-based organisations: Bridging the sacred and the secular*. New York: Springer.

Deshapurogathikk Onnikkanam: Maraman Ecumenical Sammelanam. (2006, February 23). *Malayala Manorama*, Trivandrum Edition.

Fulton, B. R., & Wood, R. L. (2017). Achieving and leveraging diversity through faith-based organizing. *Religion and Progressive Activism*, 29–55.

Hefferan, T., Adkins, J., & Occhipinti, L. (2009). Faith-based organisations, neoliberalism, and development: An introduction. In T. Hefferan, J. Adkins, & L. Occhipinti (Eds.), *Bridging the gaps: Faith-based organisations, neoliberalism, and development* (pp. 1–34). Lanham: Lexington Books.

Hefner, R. W. (1998). Multiple modernities: Christianity, Islam, and Hinduism in a globalizing age. *Annual Review of Anthropology*, *27*(1), 83–104.

Hovland, I. (2008). Who's afraid of religion? Tensions between 'mission' and 'development' in the norwegian mission society. In *Development, civil society and faith-based organisations* (pp. 171–186). London: Palgrave Macmillan.

Jodhka, S. S., & Bora, P. (2012). In the name of development: Mapping 'Faith-Based Organisations' in Maharashtra. *Economic and Political Weekly*, *47*(1), 77–85.

Mahajan, G., & Jodhka, S. (2012). Religions, democracy and governance: Spaces for the Marginalised in Contemporary India, *Economic and Political Weekly*, *47*(1), 45–52.

Menon, N. (2007). Living with secularism. In A. D. Needham & R. S. Rajan (Eds.), *The crisis of secularism in India* (pp. 118–141). Durham: Duke University Press and Ranikhet: Permanent Black.

Meynen, W. (1989). Fisheries development, resources depletion and political mobilisation in Kerala: The problem of alternatives. *Development and Change*, *20*(4), 735–770.

Occhipinti, L. (2015). Faith based organisations and the neoliberal state: Creating resistance in Northern Argentina. *Journal of International and Global Studies Volume*, *7*(1), 2.

Panikkar, K. N. (2009, September 7). Religion in the public sphere. *The Hindu*. Retrieved from https://www.thehindu.com/opinion/lead/Religion-in-the-public-sphere/article16879816.ece (accessed 27 June 2017).

Patel, S. (2010). Seva, sangathanas and gurus: Service and making of the Hindu Nation. In G. Mahajan & S. S. Jodhka (Eds.), *Religion, community and development: Changing contours of politics and policy in India* (pp. 102–128). New Delhi: Routledge.

Prakash, A. (2014). Social collectives, political mobilisation and the local state: A view from the field. *Journal of South Asian Studies*, *2*(1), 63–81.

Puthenveed, F. (1985). *The theology of the fishermen.* Quilon, Kerala: The Department of Psychology, F.M.N College.

Radhakrishnan, M. G. (1997, November 10). Thomas Kocherry takes fight for fishermen's rights from Kerala backwaters to national stage. *India Today.* Retrieved from www.indiatoday.in/magazine/environment/story/19971110-thomas-kocherry-takes-fight-for-fishermens-rights-from-kerala-backwaters-to-national-stage-832155-1997-11-10 (accessed 12 January 2018).

Rao, U. (2016). Media Hinduism. In W. Sweetman & A. Malik (Eds.), *Hinduism in India: Modern and contemporary movements* (pp. 123–140). New Delhi: SAGE Publications India.

Reeves, P., Pokrant, B., & McGuire, J. (1997). *The right to the sea: The struggle of artisanal fishers in Kerala since 1980.* Kent Street: South Asia Research Unit, Curtin University of Technology.

Roshan, M. (2011). Fishworkers movement in Kerala, India. Retrieved July 1, 2011, from: *ritimo InforméEs? Transformez*: www.ritimo.org/Fishworkers-Movement-in-Kerala-India

Sinha, S. (2012). Transnationality and the Indian fishworkers' movement, 1960s – 2000. *Journal of Agrarian Change, 12*(2–3), 364–389.

Sundar, A. (2012). Thinking beyond secularism: The catholic church and political practice in rural south India. *South Asia Multidisciplinary Academic Journal, 6*, 1–21 (Free standing article).

Sunder, A. (2017, November 17). Fishing community protest against ocean grabbing highlights growing need to protect blue economy. *The Wire.* Retrieved November, 11, 2018, from https://thewire.in/197853/fishing-community-protest-ocean-grabbing-highlights-growing-need-protect-blue-economy/

Todd Nicholas Fuist, R. B. (2017). Religion and progressive activism- introducing and mapping the field. In R. Braunstein, T. N. Fuist, & R. H. Williams (Eds.), *Religion and progressive activism: New stories about faith and politics.* New York: NYU Press.

Tomalin, E. (2013). *Religions and development.* London and New York: Routledge.

Varghese, G. K. (2014, April 04). A Hindutva variant of neo-liberalism. *The Hindu.* Retrieved June 15, 2018, from www.thehindu.com/opinion/lead/a-hindutva-variant-of- neoliberalism/article5868196.ece

Vijayan, A. J. (2017). Vizhinjam Padhadhikethireyulla Niyama Porattangal. (Personal Interview with Raoof, Abdul, 2 February).

Vizhijnam. (2015, August 13). Matsya Thozilalikalude Aashanka Pariharikkanam. *Malayala Manorama.*

Vizhinjam Padhathikkethire Thiruvanathapuram Latheen Athiroopathayude Idaya Lekhanam. (2015, August 1). *Malayala Manorama.*

Vizhinjam Thuramugham Matsya Thozhilalikale Sarkar Vanjichu. (2006, November 1). *Deshabhimani.* Trivandrum.

Williams, A., Cloke, P., & Thomas, S. (2012). Co-constituting neoliberalism: Faith-based organisations, co-option, and resistance in the UK. *Environment and Planning A, 44*(6), 1479–1501.

INDEX

Note: Page numbers in **bold** refer to tables, and those in *italics* refer to figures

For Product Safety Concerns and Information please contact our EU
representative GPSR@taylorandfrancis.com
Taylor & Francis Verlag GmbH, Kaufingerstraße 24, 80331 München, Germany

www.ingramcontent.com/pod-product-compliance
Ingram Content Group UK Ltd.
Pitfield, Milton Keynes, MK11 3LW, UK
UKHW021722240926
475702UK00029BA/1802
* 9 7 8 1 0 3 2 1 7 7 0 9 0 *